Transcontinental Silk Road Strategies

This book analyzes initiatives and concepts initiated by China, Japan and South Korea (the Republic of Korea) toward Central Asia to ascertain their impact on regionalism and regional cooperation in Central Asia.

Using the case study of Uzbekistan, the book focuses on the formation of the discourse of engagement with the region of Central Asia through the notion of the Silk Road narrative. The author puts forward the prospects for engagement and cooperation in the region by analyzing initiatives such as the Eurasian/Silk Road Diplomacy of Japan of 1997, the Shanghai Process by China, the Korean corporate offensive, and other so-called Silk Road initiatives such as One Belt One Road (OBOR) or the Belt and Road Initiative (BRI). The book argues that material factors and interests of these states are not the only motivations for engagement with Central Asia. The author suggests that cultural environment and identity act as additional behavioral incentives for the states' cooperation as these factors create a space for actors in global politics. The book deconstructs Chinese narratives and foreign policy toward smaller states and presents a more balanced account of Central Asian international relations by taking into account Japanese and South Korean approaches to Central Asia.

This book encourages wider theoretical discussions of Central Asian–specific forms of cooperation and relationships. It provides a timely analysis of Central Asian international relations and is a helpful reference for researchers and students in the fields of international relations, security studies, Asian politics, global politics, Central Asian Studies and Silk Road studies.

Timur Dadabaev is a Professor and Director of the Special Program for Japanese and Eurasian Studies at the Graduate School of Social Sciences and Humanities, University of Tsukuba, Japan. His latest publications include *Identity and Memory in Post-Soviet Central Asia* (Routledge, 2015) and *Japan in Central Asia* (Palgrave, 2016).

Routledge Contemporary Asia Series

For more information about this series, please visit: www.routledge.com/Routledge-Contemporary-Asia-Series/book-series/SE0794

Transcontinental Silk Road Strategies

Comparing China, Japan and South Korea in Uzbekistan

Timur Dadabaev

Routledge
Taylor & Francis Group

LONDON AND NEW YORK

First published 2019
by Routledge
2 Park Square, Milton Park, Abingdon, Oxon OX14 4RN

and by Routledge
52 Vanderbilt Avenue, New York, NY 10017

Routledge is an imprint of the Taylor & Francis Group, an informa business

First issued in paperback 2021

British Library Cataloguing-in-Publication Data
A catalogue record for this book is available from the British Library

Library of Congress Cataloging-in-Publication Data
A catalog record has been requested for this book

ISBN: 978-0-367-20673-4 (hbk)
ISBN: 978-1-03-209198-3 (pbk)
ISBN: 978-0-429-26282-1 (ebk)

Typeset in Times New Roman
by Wearset Ltd, Boldon, Tyne and Wear

Contents

Figures

Tables

Acknowledgments

The current study into the notion of Silk Road as a foreign policy construct has been supported by a great number of individuals and institutions.

First, a great part of the findings in this study were made during research supported by a generous research grant for international collaborative research from the Japanese Government (The Ministry of Education, Culture, Sports, Science and Technology (MEXT) (15KK0108) as well as the grant for the study of "The political impact of reconstructing Silk Road" (the leader of the project is Associate Professor Masuo Chisako with the grant number 90465386). In addition, part of this study has been completed under the grant for Comparative Study of Migration and Regional Governance (lead researcher Associate Professor Akashi Junichi, grant number 17H04543).

Second, I am grateful to Prof. Alexander Cooley and the Harriman Institute at Columbia for hosting me for 2018–2019 as a visiting scholar, during which time I had an opportunity to complete this manuscript. I am also grateful to Dr. Satu Limae and the officers of the East West Center in Washington DC for offering me their full support and facilitating my stay in Washington under the Asia Fellowship Award to undertake interviews, present my findings and disseminate these through a number of EWC's publications. I am also indebted to Prof. Marlene Laruelle for kindly facilitating dissemination of parts of this project through the Central Asia Program at the George Washington University.

I would also like to thank my family and great number of friends who are not named here individually due to the huge space this would require. Without their support and encouragement, this piece would not have been published.

1 Discursive power of "Silk Road" in China, Japan and South Korea's foreign policies toward Central Asia

Many states have used the rhetoric of reviving the Silk Road to imply closer engagement with the Central Asia (CA) region and its eventual integration into the network of economic ties. Such rhetoric is exemplified by Japan's Eurasian/ Silk Road Diplomacy, which emerged as early as 1997 under Prime Minister (PM) Hashimoto's administration,[1] South Korea's 2009–2013 "Silk Road"- related initiatives and the much discussed Chinese "One Belt, One Road" (OBOR) or "Belt and Road" initiative (BRI).[2] Alternatives to the Silk Road scheme have also been proposed by powerful states like Russia in the form of a Eurasian Economic Community and the Eurasian Union.

China, Japan and South Korea have regarded CA as a new, and perhaps the last, Asian frontier in their foreign policies over the last several decades after the collapse of the Soviet Union.[3] For these states, CA represented an area where they had not been previously active. In addition, their foreign policies in this region, at least initially, did not have any particular goals and final objectives but rather were focused on resolving the problems and issues left as a legacy of CA's Soviet past.

China has regarded CA as an area that presented it with more challenges that it had to address in the aftermath of the Soviet collapse. Multiplying the number of actors meant that instead of a unitary Soviet state, China now had to deal with several states to resolve its territorial disputes. In addition, the fact that these states were Muslim and largely Turkic presented another challenge – preventing these states from assisting those in China regarded as separatists and Islamic terrorists in Xinjiang. Thus, China regarded this region as a land of challenges rather than opportunities. With the success of the Shanghai Process and the establishment of the Shanghai Cooperation Organization (SCO), China has come to increasingly regard this region as an opportunity-generating area, a change which eventually led to the incorporation of CA into the Belt and Road Initiative (BRI).

The Japanese government has also found the task of integrating this region into its foreign policy problematic for several reasons. Its initial efforts to estab- lish a Japanese engagement strategy in CA were launched by PM Hashimoto Ryutaro in the form of Eurasian (Silk Road) diplomacy (1997–2004), which referred to engaging Russia, China and CA.[4] Hashimoto hoped to bring the nations of the former Soviet Union into a network of interdependence by establishing a largely economic and, to a degree, political Japanese presence in Eurasia and by facilitating Japanese participation in resource exploration.[5] Although Japan's presence has been supported by economic and humanitarian assistance as well as various important initiatives (such as the initiation of the CA plus Japan dialogue of 2004, PM Koizumi Junichiro's CA visit of 2006, and Foreign Minister (FM) Aso's 2006 "Arc of Freedom and Prosperity" speech) and the 2015 PM Abe visit to CA, its strategy of engaging these states has faced a number of challenges, including its geographic distance from the region, limited corporate penetration of CA and lack of a clearly defined strategy.[6]

South Korea is similarly distant from CA and lacks transportation infrastructure to and from markets in CA. Its infrastructure toward the Eurasian region is also dependent on improving its ties with North Korea. However, the South Korean "Silk Road" rhetoric is somewhat more practical than the Japanese rhetoric. South Korean corporate economic interests were present and more active in the region beginning in the early 1990s, with Daewoo building a major car manufacturing plant in Uzbekistan; in addition, a large number of assembling and manufacturing facilities were built under the Samsung and LG brands in Uzbekistan and Kazakhstan to locally produce electronics parts. In the early 1990s, South Korea was among the leading investors in this region (Daewoo car plant, the Daewoo Unitel, Kabool Textiles, etc.).[7] These corporate advances later translated into President Roh Moo-hyun's Comprehensive Central Asian Initiative of 2006, the Korea plus Central Asia forum creation of 2007, Lee Myung-bak's New Asia Initiative of 2009 and Park Geun-hye's Eurasia Initiative of 2013 as well as a state visit by South Korean president Moon Jae-in in April 18–21, 2019 to Uzbekistan.

While Japan, China and South Korea appear to face similar limitations and challenges in approaching CA, it is unclear how different or similar their agendas are with regard to economic cooperation with this "Silk Road" region. Additionally, little is known about the operationalization of their economic cooperation agenda setting given the conditions of the change of leadership in some of the states of CA.

This study focuses on the formation of this discourse of engagement with the CA region by China, South Korea and Japan and raises the following questions: What are the discursive approaches that facilitate the most effective ways of engaging CA states by China, Japan and South Korea? What are the principles that have detrimental effects on the successes and failures of the discursive engagements of China, Japan and South Korea? What are the projects that these states plan to implement in the region and how do they relate to and differ from each other? The primary objective of this study is to address these questions and to stimulate debate among both academics and policy makers on the formats of engagement and cooperation in Eurasia.

Thus, by seeking answers to these questions, this study attempts to achieve the following objectives. First, the study analyzes the impact of the various concepts and initiatives undertaken by China, Japan and South Korea (Eurasian/Silk

Road Diplomacy of Japan of 1997, Shanghai Process by China, Korean corporate offensive, Chinese so-called Silk Road – One Belt One Road, OBOR or the BRI – to name a few) on the prospects for regionalism and regional cooperation in Central Asia in conceptual and theoretical realms. Second, while this study highlights the nuances of China's initiatives and their Japanese and South Korean (Republic of Korea, ROK) counterparts, its primary goal is to go beyond the simple empirical consideration of the facts to identify the implications of this initiative for wider theoretical discussions of CA-specific forms of cooperation rooted in the nexus of relationships found in this region. While the coverage of this study in the first few chapters relates to relations of these three states with all CA states, the primary focus of this study is to trace the discourses of Chinese, Japanese and South Korean foreign policies toward Uzbekistan.

The reasons for choosing Uzbekistan are multiple. First, Uzbekistan is the largest CA state in terms of demographic composition. Uzbekistan's stability and development have a detrimental impact on the sustainability of regional development in the CA region. Second, Uzbekistan is one of the few CA states that attempts to build an equal balance in its relations with Russia, the EU and Asian states by always emphasizing that China, Japan and Korea are its strategically important partners. In comparison to it, other CA states such as Kazakhstan, Kyrgyzstan or Tajikistan tend to over-depend on Russia (through membership in the Eurasian Economic Community and Eurasian Economic Union) or on China (through a huge share of debt). In contrast, Uzbekistan has pursued a foreign policy aimed at limiting its dependence on international actors through eschewing military alliances and balancing relations with larger powers. In this sense, Uzbekistan is uniquely positioned to be analyzed as the country that does not necessarily favor one of these states as its most important economic partner while attempting to define the importance of each of these states for its economy. Third, Uzbekistan is the state that is currently transitioning from the previous president Islam Karimov's dictatorial type of government to openness, embracing foreign engagements with various countries in the post-Karimov era.[8] Thus, the case of Uzbekistan demonstrates the challenges of CA states in their engagements with their more powerful Asian counterparts. In addition, an analysis of the post-Karimov-era economic engagements of Uzbekistan with East Asian countries demonstrates the elements of continuity and change in its foreign policy, thereby offering insights into its behavior for the foreseeable future. Finally, the choice of Uzbekistan is unique from the perspective of inquiring into China, Japan and Korea's foreign policy. None of these states enjoyed necessarily friendly relations with CA in general or with Uzbekistan in particular prior to the collapse of the Soviet Union, because these states constituted the Soviet space. Since the collapse of the Soviet Union, these countries have represented the new political and economic frontier where China, Japan and Korea can construct their relations in the conditions of a changing international order and the changing nature of their international standings. All of these states launched their initiatives in Uzbekistan to start in 1991. Thus, their starting positions in this region were somewhat similar. China has now grown into the second largest world economy; thus, it is adapting to the necessity of dealing with smaller neighbors such as Uzbekistan. Japan and Korea are also in the process of adapting their behaviors to the conditions when their economic power has faded compared to China while they still see the need to expand their outreach into the CA region in search of new standings and the opportunities that such standings entail. Japan is in search of a new place and role in this Japan-friendly region where there is an articulated expectation of a larger Japanese presence as demonstrated by various polls. In the case of Korea, it has invested heavily through its corporate penetration in CA and is thus interested in expanding its economic presence. It is especially important for its presence in Uzbekistan where it enjoys support from the government after the successful visit of the Uzbek president to South Korea in 2017. In addition, both countries are engaged in region-building efforts with the CA plus Japan initiative and the Korea plus CA forum. While these are not counterposed to the Chinese SCO and BRI schemes, they represent Japanese and Korean efforts to present an alternative "other" to CA states. In this sense, the analysis of road maps of cooperation provides insights into foreign policy behaviors and factors important for efficient cooperation as well as shedding light on the possible challenges in engagements with CA states, exemplified by the case of Uzbekistan with China, Japan and Korea.

Enquiring into Chinese, Japanese and South Korean intentions

This study throughout its eight chapters employs a three-fold methodological approach. First, it is an exercise in the theoretical re-consideration of the Chinese, Japanese, South Korean and alternative cooperation schemes of the Silk Road to which CA states are exposed. It then considers the most significant alternatives to the Silk Road by criticizing old clichés of engagement and contestation. More specifically, the study questions the narrative of domination and conquest of CA by economically larger states. Instead, this study argues that the discursive practice of using the various terms (as exemplified by "Silk Road") has different connotations and meanings in different (Chinese, Japanese and South Korean) settings. Comparing the usages of such seemingly similar terms like "Silk Road" offers an opportunity to identify factors that are important for successful cooperation among these states. In this way, this study aims to promote and extend current international relations efforts to place CA within appropriate comparative frameworks.

Second, this study tests and partially deconstructs the notions of the "Silk Road" and its "others" to determine whether the "Silk Road" represents a neocolonial construct or offers decolonizing perspectives that lead to new identity construction in the CA region. It does so through analyses of the initiatives, statements, agreements and speeches of political leadership.

This study primarily builds its assumptions on the expressions of political discourses exemplified by speeches of the president of PRC and the foreign minister, along with statements by the Shanghai Cooperation Organization.[9]

To operationalize discursive coverage, this study uses the speeches and initiatives of heads of states and governments, including both FMs and certain ministries and agencies. In the case of Japan, the main actors in the discourse are the PM, the Ministry of Foreign Affairs (MOFA), its affiliated agency the Japan International Cooperation Agency (JICA), the Ministry of Finance and others. In addition, academic discourse partly related to the official narrative in this study sometimes influences decision making by shaping the environment for certain decisions. In the case of China, the main actors who produce and popularize the discourses on CA are the president (chairman) of the People's Republic of China, the Foreign Ministry and local municipalities involved in interactions with their CA counterparts. In addition, the statements within the SCO, OBOR/BRI and "Central Asia plus Japan" (CAJ) initiatives are treated as indicative of Chinese and Japanese discourses regarding CA. In the case of South Korea, these are initiatives by the president, prime ministers and foreign ministers (President Roh Moo-hyun's Comprehensive Central Asian Initiative of 2006, the Korea plus Central Asia forum creation of 2007, Lee Myung-bak's New Asia Initiative of 2009 and Park Geun-hye's Eurasia Initiative of 2013, etc.) as well as discourses provided by the Korean International Cooperation Agency (KOICA).

And third, this study also recognizes that while discourses on the intentions of various powers in engaging CA states have been analyzed on multiple occasions, few studies, if any, consider the particular projects these states plan and analyze their reasoning and implementation. This study notes that while speeches and statements of presidents, foreign ministers and policy officials inform our understanding of the relations between CA and East Asian states, there is a need to look into the practical steps taken by these states. This conceptual gap is addressed in the following chapters of this study that focus on an analysis of economic cooperation road maps and demonstrate how these politically articulated intentions materialize in the practical realm. That is not to say that economic road maps are necessarily realizable plans. However, they are the most tangible plans and the closest to practical outcomes of governments' articulated intentions and discourses.

Analysis based on these three parts of the framework demonstrates not only the goals and principles according to which both China and Japan operate in this region but also that their declared interests and actions do not conflict, demonstrating that the Chinese–Japanese rivalry is largely a matter of myth and perception that is spread for various reasons depending on the actor. Such perceptions in Russia, for instance, are often an indication of Russia's insecurity or a perception that Russia's "backyard" is being contested. Such views among CA elites often have the purpose of attracting more attention to the region from other players by intentionally emphasizing Chinese–Japanese rivalry in an attempt to demonstrate the region's attractiveness. It can also be interpreted as a bargaining strategy by CA states for larger investments from Turkey, India, Europe or other players. Media discourses that fuel such images of rivalry in CA have the commercial purpose of "selling" the story to a wider audience.

The selection of case studies (China's OBOR/BRI; Japan's Silk Road Diplomacy and the initiatives under the PM Koizumi and PM Abe administrations; and South Korean president Roh Moo-hyun's Comprehensive Central Asian Initiative of 2006, Lee Myung-bak's New Asia Initiative of 2009 and Park Geun-hye's Eurasia Initiative of 2013) herein corresponds to geographic criteria and is also designed to reflect a variety of different strategies of cooperation schemes.

This study also recognizes that the discourses generated from the aforementioned speeches, while overlapping at times, often target different audiences, with the English-language ones largely produced for international consumption, Russian ones often aimed at audiences in Central Asia and Russia, Chinese (Mandarin Chinese) ones for domestic audiences in China and the Japanese language ones produced for the Japanese public. In most cases, this study uses these documents as published in English and refers to the original (Japanese, Chinese and Russian language) documents when there is no English version.

The general theoretical framework, detailed in the arguments of this study below and in each chapter, used to explain the engagement strategies of China, Japan and South Korea and CA/Uzbekistan's responses analyzed in this study is a constructivist one. It claims that these strategies are socially constructed and are constantly undergoing reformulation and reshaping. At the same time, each case study considered in this study has its own unique specificities. Thus, the analysis is conducted by considering cooperation within OBOR/BRI initiatives as emphasizing pragmatism and functionalist goals in contrast to the integration processes in the Commonwealth of Independent States (CIS) after the dissolution of the USSR. This study then considers similar attempts by Japan and South Korea to engage post-Soviet CA through constructivist lenses.

Depictions of CA in international politics

Many studies have been published in recent years focusing on the foreign policy of various powers in Asia. However, these studies tend to focus primarily on the countries and areas in Asia that have historically received extensive attention, including China, Japan and South Korea in East Asia. Few studies go beyond traditionally covered areas to focus on parts of Asia that, while becoming central to various international engagements, remain overlooked. One such example of an area not paid due attention in the literature on comparative aspects of foreign engagements of Japan, China and Korea is what can be referred is the last "new frontier" in Asia – Central Asia (CA).[10]

The post-Soviet CA region – consisting of the five stans of Uzbekistan, Kazakhstan, Kyrgyzstan, Tajikistan and Turkmenistan – has remained marginal for Asian scholars for a number of reasons. First, this region has often been associated with the geopolitically determined larger Eurasian region consisting of Russia and other post-Soviet constituencies. Thus, for many scholars in international relations (IR), this region has been approached through the analysis of Russian and post-Soviet policies, while the Asian angle of CA states' interactions has been shadowed and to some extent hijacked by Russia-related scholarship.[11] Second, those few studies that did pay attention to the CA states' interactions with Asian powerhouses, in comparative perspective, tended to focus on these states' participations in the Shanghai Cooperation Organization (SCO) or their foreign policies related to the recently announced Belt and Road Initiative (BRI).[12] Thus, once again, the framing of the CA region's coverage within the Asian political space has been hijacked by the attention paid to China-related initiatives which are often justified by the rise of China and its global and regional economic influence.[13] Third, those studies that intended to cover CA states' engagements with Asian countries frequently focused on individual case studies of CA–China, CA–Japan or engagements between this region and South Korea.[14] Very few studies, if any, have attempted to consider the mutual importance of CA states and powerful Asian countries in such interactions.[15] In addition, differences and similarities in Chinese, Japanese and Korean interactions have rarely been compared.[16] However, any conclusions on the role and significance of the Asian vector in foreign policies for CA states are difficult to make without an empirically grounded comparison of CA interactions with the most important and active states in this region: China, Japan and South Korea.

Discourses of engagement and contestation in CA

In the 28 years since the independence of the CA states, the issues of how to form and shape cooperation between and among these states have been some of the most discussed issues in regional international relations. These include but are not limited to discussions on the formation of the CIS, Central Asian cooperation, the Shanghai Five, Shanghai Cooperation Organization and Eurasian Union. Interestingly, many of these schemes have been considered from two main positions.

The first position is the notion of engagement. Integration of their economies into the structure of the international economy has been of crucial importance to CA states. Therefore, most schemes offered to these states represented methods of engaging them and forming closer relations among their economies and with countries outside the region. To a great extent, these attempts represented decolonization efforts by CA states and their counterparts through the diversification of their infrastructure and trade partners and an effort to attract foreign investors to the region. The degree of success in these areas depended on the degree of participating states' commitment and motivation. The outcomes of these efforts in the areas mentioned above depended on the efforts of CA states and on the changing international environment. Although most of these schemes resonated with the intentions of the CA states, they did not produce many tangible outcomes. On the contrary, the concept of cooperation has been corrupted by the great number of these cooperation schemes and the lack of outcomes produced. Such a lack of efficiency led to claims that most of these schemes were initiated as a result of attempts to dominate and control the CA region and to access regional resources. As a consequence, interpretation and narration about these schemes frequently developed along the lines of conspiracy theories reproducing the discourse of CA being contested and eventually turning into a battleground for neo-colonization and a new "frontier" to be "conquered" by "big players". To some extent, CA states themselves are also responsible for the regeneration and reproduction of these discourses, as they often misinterpret the importance of engagement with the international community as a form of contestation. For many of these CA governments, engaging with larger states inevitably results in contestation, which they attempt to use as a bargaining tool to receive larger amounts of developmental aid. However, the discourse of contestation in the sense of rivalry for dominance in the CA region does not reflect the intentions of other states. Rather, the contestation witnessed in the region is about the types of engagements and models of interactions that are employed by various countries.

In this sense, the discourse of contestation for the CA region has multiple aspects. Interestingly, the discourse of the creation of a common new Eurasia has been approached differently by different states. Some conceptualized the issue of Eurasian cooperation from a conventional integration perspective, suggesting that such elements as historical ties and common linguistic and civilizational features constitute the basis of and an asset for cooperation. In Russian constructs, these are mostly related to "common belonging" to the former Soviet space. References were also made to the notions of Eurasianism and shared geography. According to these notions, geography and common history produce a common identity, which then leads to the commonality of approaches to conceptualizing cooperation. Under this view, the notion of common Eurasian norms was frequently emphasized as a part of a shared value system that helped CA states integrate with Russia and other post-Soviet constituencies. Among these norms and values, the emphasis on collective/group rights over individual human rights and the paternalistic role of the government in the social contract in these societies have frequently served as examples of ideas that unify the approaches of many states in post-Soviet Eurasia. However, the approach that relied primarily on civilizational, ethnic and historical commonalities did not sustain the motivation of participating states. The best examples to illustrate such schemes are CIS and the Central Asian Cooperation. With time, these schemes were more associated with attempts by Russia to dominate this region or, in the case of CA Cooperation, with contestation between Kazakhstan and Uzbekistan for dominance and leadership in the region.

After the inefficiency of CIS became apparent, the states participating in such schemes began to pay more attention to sustaining the motivation of member states of cooperation schemes. The Chinese schemes were largely a response to the problems existing in the relations among member states (Shanghai Five).[17] This problem-solving approach toward engagement has several connotations. It serves the functional purpose of resolving particular problems and, at the same time, is only possible because parties develop common approaches, norms and mutual trust within this problem-solving process, thereby leading to common norm creation. In this sense, the Chinese attempted to create schemes that are not necessarily (initially at least) based on the notion of "common belonging" but have more of a problem-solving, reactive nature, such as SCO, later developed into the OBOR/BRI and related Silk Road transportation network that exemplified this pattern. They can also be considered as a response to the deficiencies of previous schemes, such as the deficient functioning of the CIS. Therefore, instead of focusing on the issues of civilizational and conventional cooperation, Chinese initiatives focused on common-approach and norm creation for practical problem solving because these targeted areas were either of a less politically sensitive nature or were related to the real needs of each CA state. Another difference in the SCO and post-SCO structures from other previous formats is that bilateral engagements are frequently designed or sometimes simply interpreted as those produced by a multilateral structure. There is also the possibility that many of these outcomes claimed to be achieved multilaterally would probably have happened through pre-existing or current bilateral processes and that the SCO simply places a supposed multilateral stamp on what is essentially agreed upon bilaterally. However, attributing these achievements with a multilateral nature legitimizes China-led SCO and post-SCO structures and strengthens the sense of "common belonging" and common "identity".

Nevertheless, when transportation networks and the ever-growing Chinese economic power began to show signs of larger Chinese corporate penetration of CA markets and appeared to benefit China more than CA counterparts, these Chinese projects became a concern to the participating states.[18] Chinese moves were subsequently balanced by the expansion of SCO and the entry into Eurasian integration schemes of smaller CA states. In turn, Russia also appears to have learned from its past mistakes. It now attempts to construct a vision of a cooperation scheme that will prioritize common identity creation through schemes that are oriented at achieving practical and feasible outcomes. These assumptions aim at increasing the efficiency of cooperation in Eurasia. Although Eurasian construction is a notion that is more frequently interpreted from the rationalist perspective, this study claims that the notion of Eurasian political space construction needs to be understood in constructivist terms. The reason for such a claim emanates from the understanding that Eurasian political space is a continuing process of construction that involves construction of discourse, practices and perceptions. It also involves the process of counterposing what Eurasia is and is not. In such a process of construction, the notions of "practicality" and "functionality" also become pieces that contribute to the process of trust- and norm-building, which inter-subjectively develop into a common identity. Thus, there is no tension between functionalist engagement and constructivist discourse in narrating Russian Eurasian engagement discourse.

Japan and South Korea show a tendency to differently conceptualize their vision of CA engagements, thus contesting Chinese and Russian regional constructs. As described in the chapters of this volume, the Japanese engagement in this region argues strongly for "open regionalism", which corresponds to Japan's distant location from the region and attempts to allow participation of non-regional states. In this sense, Japan contests the notion that the CA region is only available to the big powers bordering CA. However, the Japanese engagements in this region are mostly done through the official development assistance (ODA) scheme conducted by the Japanese government, with limited participation of corporate interests. Although the Japanese government claims that such limited Japanese corporate participation is a competitive advantage of Japan because of the absence of Japanese self-interest in CA engagements, such a claim of "altruism" in CA engagement is part of the discursive construction of competitive advantage in respect to other big players, such as China and Russia.

Another alternative to the Chinese, Japanese and Russian modes of engagement is South Korean diplomacy, which is rooted in President Roh Moo-hyun's Comprehensive Central Asian Initiative of 2006 and Lee Myung-bak's New Asia Initiative of 2009. Prior to the announcement of this initiative, PM Han Seung-soo conducted a mission to the Caucasus and CA, during which he visited Kazakhstan, Turkmenistan and Uzbekistan. After the announcement of the "Silk Road Diplomacy" initiative, President Lee upgraded the status of relations with Kazakhstan and Uzbekistan to the level of strategic partnerships and provided the framework for governmental support of Korean corporate penetration into this region.

From SCO to OBOR/BRI: Spill-over or a shift?

The primary appeal of the SCO for its members is the decolonization opportunities that this organization generates for its participating states.[19] As an international organization, the SCO functions by embracing the principles of non-interference in the internal affairs of its states and respect for the sovereignty and developmental choices of each member state.[20] Explicit references to this respect and consideration for mutual concerns within the SCO charter as well as the implementation of cooperation based on the notion of the Shanghai spirit further aid the organization's anti-colonial rhetoric.[21] Therefore, the Shanghai spirit as a value system that is applied to cooperation serves the role of an anti-colonial safeguard, although there are several reservations about neo-colonial tendencies within SCO, which the next section explores. This anti-colonial stance is especially important for those CA states with transport and energy-related infrastructure constructed in a way as to connect them to external markets, primarily through Russia. The Soviet Union attempted to construct a system to facilitate economic modernization in this region under the model of Soviet modernity. It also aimed to increase the degree of interdependence between Soviet constituencies, thereby reinforcing the notion of Sovietness within the identity of CA republics. However, post-Soviet governments regard such Soviet policies with a degree of skepticism, considering them a way to colonize CA republics. According to the prevailing CA political discourse regarding Soviet-era policies, the Soviet leadership attempted to control the vast energy resources of these republics, providing only a tiny portion of the revenues received from their exports to CA republics.

The SCO and China's proactive policy to facilitate various transportation and energy infrastructure projects are generally welcomed by CA governments, as these projects offer them more means to deliver their products to international markets while limiting the dominating influence of Russia. This organization is also sometimes considered to be a proper substitute to certain functions of the Collective Security Treaty Organization, within which Russia plays a major role.[22] Some CA governments consider China to be a proper safeguard against not only Russia but also the US (perceived or real) in its efforts to widen its sphere of influence in CA. These states support the US's efforts to stabilize Afghanistan because these efforts also contribute to the stability of the CA region in general. However, these states strongly fear that a US presence in this region would also entail heavy criticism of regional governments' human rights records and "political mentorship", as seen during the government take-over in Kyrgyzstan and the US criticism of the Karimov administration during Andijan events in Uzbekistan in 2005.

Therefore, the SCO's strong stance against the foreign military presence in the CA region are welcomed by regional states.[23] Russia still maintains bases in Tajikistan and Kyrgyzstan and important military facilities in Kazakhstan and thus does not necessarily share this anti-imperial SCO stance. However, it accepts this stance because Russia possesses alternative means of influencing the CA region, as exemplified by the significant number of CA labor migrants in Russia whom Russia can potentially instrumentalize to influence these countries, without a need for military action.

In addition to the anti-imperial stance mentioned above, the Chinese model of generating economic growth is an appealing point for many CA states in their participation in SCO structures. The developmental state model of economic development, which aligns with significant government control over various aspects of public life under the conditions of the unstable political and economic systems in CA states, is a model of modernization that is radically different from Western models of free liberal democracy.

There is certainly a danger that Indian and Pakistani membership (with a population many times that of the combined population of CA) in practice can shift the agenda away from CA and from the smaller states' priorities. However, if safeguards against such drastic shifts are put in place, successes, principles and lessons from the SCO construction process can be utilized for the benefit of building a new engagement in this region; this is the strategy adopted by the OBOR/BRI initiative announced by China in 2013.

The new Chinese OBOR/BRI distinguishes itself from the SCO in its wider geographic coverage as well as in the issues that it covers. There were certain attempts to reform the SCO to expand into economic areas but, by 2012, it was obvious that such "maturing" of the SCO was paralyzed by the different preferences of Russia and China. Each country had its own distinct vision of free trade zones, financial institutions and priority issues. China emphasized the creation of a free trade zone within the SCO and the creation of an SCO bank and related economic activities. Russia, however, preferred the SCO to focus on security-related issues and prioritized constructing its own Eurasian Economic Community with post-Soviet constituencies, which is eventually supposed to mature into the Eurasian Union. To some extent, such inexplicit rivalry between the visions espoused by Russia and China facilitated an international environment for the Chinese shift from the SCO to OBOR/BRI. In this sense, the influential role of Russia in shaping and largely constraining the Chinese from pursuing their vision for the development of the SCO needs to be emphasized. In other words, the development of OBOR/BRI occurred, in a certain sense, at the expense of the SCO, which steadily loses functionality in terms of project development and implementation as OBOR/BRI expands and dominates the Chinese political agenda.

The "problem-solving" principle, which was largely inherited by OBOR/BRI from the SCO, could serve as one of the norms within the innovative approach to cooperation. To succeed, in stark contrast to the SCO, OBOR/BRI would also require bottom-up "soft regionalism" involvement and not simply the top-down presidentially determined projects currently seen within the SCO. Partly in response to these challenges, there are already signs of engagements in this region that have developed not only between governments but also covering region-to-region cooperation. For example, the Chinese government has encouraged provincial leadership to seek opportunities for Chinese provinces in CA. Similarly, CA states, like Uzbekistan under the new Mirziyoyev administration, attempt to develop region-to-region ties to encourage investments, an approach which has already produced several noticeable outcomes.

OBOR/BRI and its Japanese/South Korean variations

Although China has extensively engaged in CA over the past decade, it aims to further expand its presence in this region by offering a transportation infrastructure project that would help China further penetrate the CA market and allow further penetration of other regions. As a part of these plans, in 2013, the Chinese government proposed the construction of the "Silk Road Economic Belt" (consisting of six economic corridors, of which the Eurasian Land Bridge, China–Central Asia–West Asia, and the "Twenty-First Century Maritime Silk Road" are relevant to CA) OBOR/BRI concept; this represents an even broader and more ambitious initiative than that represented by the SCO.[24] The Asian Infrastructure Investment Bank (AIIB) that was established as an alternative to regional and global financial institutions, such as the Asian Development Bank or the World Bank (in partnership with BRICS – Brazil, Russia, India, China and South Africa – New Development Bank and Silk Road Fund, SCO interbank association), will serve as the financing arm of the OBOR/BRI.[25] In addition to the economic benefits of this project, the Chinese government aims to address the lack of trust from some of its smaller neighbors by offering infrastructure development, which sends a message to CA states that China is genuinely interested in contributing to their development.[26] This is supposed to further strengthen the "soft-power" potential of China in Central Asia by offering a non-coercive, non-military (non-security focused) approach.

The OBOR/BRI initiative represents an increasing effort to shape positive Chinese involvement in CA by constructing a sense of common belonging based on pragmatism and functionality. On many occasions, Chinese officials have emphasized that OBOR/BRI is not an "expansion of China" but rather a benefit for all parties.[27] However, both CA politicians and the public have often expressed the view that Chinese initiatives under the umbrella of SCO and OBOR/BRI largely benefit China while making little if any contribution to the long-term development of this region.[28] These views have sustained the image of the continuous expansion of China, have led to anti-Chinese protests in Tajikistan, Kyrgyzstan and Kazakhstan, and have potentially demotivated further engagement of CA states with China.[29]

Currently, OBOR/BRI initiatives involve such economic components as developing energy resource transportation and railroad and highway infrastructure. However, the effects of these projects on strengthening the growth-generating potential of CA states is frequently questioned because previous initiatives have had little effect in facilitating economic development in the region.[30] In this sense, the success of OBOR/BRI largely depends on how China and other regional states will learn from the successes and failures of the SCO and other engagement schemes in this region.

In terms of Japanese involvement, similar calls for a more mutually beneficial structure of relations were heard from CA countries during the recent visit of Japan's PM to Central Asia on October 22–28, 2015.[31] The visit of PM Abe to Central Asia can be termed historic because it was the first ever visit of the Japanese PM to all five Central Asian states. PM Koizumi visited Central Asia in 2006, but his visit was limited to Uzbekistan and Kazakhstan. PM Abe's visit builds upon previous Japanese engagement strategies exemplified by PM Hashimoto's Eurasian Diplomacy of 1997, the Obuchi Mission of 1998, FM Kawaguchi's visit of 2004 when the "Central Asia plus Japan" forum was established and PM Koizumi's visit in 2006 as mentioned above. Japanese experts generally attempt to differentiate the Japanese engagement in CA from the Chinese one by arguing that the Chinese "assistance" to CA is largely a gesture of goodwill like a "firework shoot" before the launch of major infrastructure-related projects.[32] For them, Japanese involvement in this region goes beyond infrastructure construction and attempts to transfer technology and knowledge.

The goals of PM Abe's visit to CA in 2015 partly confirm the description above and can be considered four-fold. First, the Japanese PM attempted to deepen and strengthen the presence of the Japanese business community in Central Asia, as exemplified by the contracts signed during the visit for the joint exploration of gas fields in Turkmenistan (Galkynysh), Uzbekistan and Kazakhstan. Such intensification of direct investments by Japanese companies with the support of the Japanese government has been encouraged by the majority of CA governments, as exemplified by the speech of the Tajik president, who explicitly emphasized the importance of a switch from humanitarian aid projects to foreign direct investment by Japanese companies.[33] Second, it was an attempt by the Japanese PM to secure orders from CA countries for Japanese corporations, as exemplified by the Japanese and Kazakh commitment to work on the construction of a nuclear plant in Kazakhstan and the Japanese and Turkmen agreement (between Turkmengas and JOGMEC, the Japan Oil, Gas and Metals National Corporation) on the construction of mineral resource processing factories in Turkmenistan. Currently, Kazakhstan is also negotiating with Russia on the possible construction of such a plant. Third, PM Abe aimed to further boost its "Cool Japan" soft-power construction initiative by supporting the construction of a Japanese university in Turkmenistan, cooperation on IT education in Tajikistan, by launching the Youth Technological Innovation Center in Uzbekistan and similar educational initiatives in Kazakhstan.[34] Fourth, Abe's visit offered further humanitarian aid to the smaller republics of Tajikistan and Kyrgyzstan for various development-related projects.

Some experts have argued that Japanese technologies and loans for the projects mentioned above are a sign of competition between China and Japan because they offer alternatives (to Chinese projects) and thus competing sources of funding.[35] Some have even suggested that "incongruent interests between the two powers already hint for the potential for a friction in the region".[36] However, there is little evidence to suggest that these loans for Japanese projects in Turkmenistan aim to affect Chinese projects. Although the field of mineral extraction coincides with Chinese interests, Japanese loans and projects in Turkmenistan do not aim to exclude China's investment there. In addition, the Japanese initiatives extend beyond energy resources to encompass human resource development, joint university and research facilities construction and human security infrastructure.[37] Additionally, former and current MOFA officials frequently attempt to define the features of "Japanese-ness" in assisting and engaging the CA region that are different from the features offered by the Chinese counterparts.[38] Therefore, it is difficult to argue that Japan directly links Japanese initiatives with the intensification of Chinese or Korean initiatives in this region. Japanese attempts to secure access for Japanese corporate and state institutions to CA may offer some alternative resources for funding and development initiatives to those offered by Chinese and Russian schemes. However, at least in the official discourse and project implementation, Japanese involvement in CA does not appear to be linked to countering Chinese or Korean initiatives because the goals and the degree of commitment to the region differ between Japan and other global players. Interestingly, there are suggestions within the expert community in Japan that OBOR/BRI initiatives in CA do not contradict with the Japanese engagements. Rather, experts suggest that Japan should use China-built and financed infrastructure for the benefit of Japanese corporate penetration.[39] As has been voiced by PM Abe, OBOR/BRI has potential; it is important that both Japan and China contribute for the peace, prosperity and resolution of international problems, and Japan is prepared to cooperate (with China) where it can.[40] At the same time, PM Abe emphasized that OBOR/BRI infrastructure schemes need to be transparent, and loans provided under this scheme should be repayable by borrowing countries without indebting these developing nations.[41] The Chinese MOFA welcomed such a Japanese stance by saying that

OBOR/BRI will become a "development platform, creating benefits for countries around the world including Japan" and responded to the Japanese PM's remarks by saying that "China is committed to establishing a set of fair, reasonable and transparent rules for international trade and investment".[42] Therefore, both countries' foreign policies are in the indirect process of mutual shaping and dialogue, which again confirms constructivist logic.

South Korea has also attempted to use opportunities created by the Chinese OBOR/BRI initiative for Korean gains. However, it should be noted that South Korean Silk Road Diplomacy was launched earlier than the articulated OBOR/BRI initiative. Thus, it represents a branding umbrella scheme for expanding South Korea's corporate interests in this region. By the time South Korea's Silk Road Diplomacy was announced, South Korea's economic presence in CA, in particular in Kazakhstan and Uzbekistan, was significant in terms of ODA assistance, direct investments and human resource development. In terms of aid disbursements, Uzbekistan ranks the highest in the region, accounting for a third of the assistance volume, while Kazakhstan is second highest, accounting for approximately 6 percent. South Korean assistance and investment programs, in addition to considering importing resources from CA to South Korea (similar to the 2008 contract to import 2,600 tons of uranium from Uzbekistan), also aim to link CA resources to international markets. In this way, they aim to establish economically sustainable production, extraction and reproduction cycles that are marketable in CA and beyond the region, bringing the benefit to the South Korean conglomerates. In particular, during President Lee's visit to CA in 2009, he secured an agreement to build a petrochemical plant in Atyrau with a budget of US$4 billion, a US$4-billion contract to construct power generation plants in Balkhash and a US$4-billion deal to participate in development of a gas field in Uzbekistan. In 2011, during another visit by President Lee, the construction of the Ustyurt gas chemical complex was launched. In addition, South Korea's Hanjin (parent company of Korean Air) received exclusive rights from Uzbekistan to develop its Navoi International Airport into an intercontinental logistic hub.

Thinking structurally about the discursive power of China, Japan and South Korea in Central Asia

After laying out the main ideas, dilemmas and approaches of this study in this chapter, the analysis is presented in seven further chapters, each detailing various aspects of China and Japan in Central Asia. This study does not limit its coverage to China and Japan but also extends to other states which have proactive standing in the region. In particular, South Korean engagement in the CA region has provided a case for comparison which frequently falls in between the Chinese and Japanese approaches and presents a unique case requiring further analysis. In some other cases, references are made to Russian initiatives of the past and present as an example of "othering" for China and Japan in their engagement in the region.

Chapter 2 presents the evolution of the Japanese, Chinese and South Korean discourses and narratives of the Silk Road from the early 1990s to the present. This chapter argues that the content and the nature of these Silk Road strategies changed with time and the international environment. Thus, this chapter claims that the notion of the Silk Road has changed from a static concept of a historical trade route into a product of social construction of a number of powerful states – strategies that are constantly shaped, imagined and re-interpreted. In this sense, the Silk Road is not a foreign policy doctrine but rather a discursive strategy of engagement that largely exists in the realm of narration. This narration is also a matter of social construction that is subject to change depending on the international environment of the country (China, Japan, Korea, etc.) that produces such narratives, context of a receiving region, the alternative narratives that compete for wider international acceptance and the country's vision of "self" and the "other" in the international context.

Chapter 3 analyzes discursive strategies of China and Japan of recent years to integrate newly emerging CA states into their internal and external policies, norms and concepts, according to which they justify both their actions in CA and CA responses to these policies. This chapter elaborates the concept that to a certain extent, the interests of China and Japan in CA are similarly focused on mineral resources and political stability. However, these countries employ different discursive strategies to frame their approaches and goals. This chapter, in line with the general argument of this study, also emphasizes that the discourse of competition for regional domination prevalent in the English-language, Russian and some Central Asian media is largely an imposition of a zero-sum vision of international relations that is not proven by any empirical evidence. On the contrary, many of the projects conducted both by China and Japan are compatible – if not supplementary – and do not necessarily imply exclusivity of interest. At the same time, China and Japan have different ways of reasoning their CA engagements, resulting in a rivalry of discourses for the "hearts and minds" of the Central Asian population.

Chapter 4 further compares Japanese and Chinese infrastructure development strategies in post-Soviet CA by detailing on the similarities and differences in the approaches of the two Asian economic powers in their infrastructure development in this region. This chapter develops several arguments with respect to the Japanese and Chinese approaches to infrastructure development in CA. This chapter argues that the analysis of various infrastructure-related projects undertaken in CA by China and Japan demonstrates that the discourse of mutually exclusive interests in China and Japan in this region is empirically unproven. Most of the Chinese engagements emphasize the creation of energy and transportation infrastructure (construction), while Japan's main areas of focus are the maintenance, modernization and rehabilitation of current infrastructure. Thus, this chapter suggests that China positions itself as CA's leading economic partner, while Japan is CA's leading assistance provider. These two roles have different implications. Furthermore, the current infrastructure engagements of Japan (from assistance to partnership) and China (from exploitation to contribution to the region) in CA demonstrate both countries' attempts to adjust and search for new international standing.

Chapter 5 zooms in on the analysis of the economic road maps for cooperation between China, Japan and South Korea, on the one hand, and Uzbekistan, as an example of CA country, on the other. This chapter focuses on the bilateral economic cooperation agenda setting of China, Japan and South Korea and their counterpart in CA, Uzbekistan. The importance of focusing on China, Japan and South Korea can be explained by these countries' roles in terms of economic contribution and official development assistance to CA in general and Uzbekistan in particular. Uzbekistan is chosen not only due to word limitations of this study but, importantly, because Uzbekistan is the demographically largest and geographically most central country of this region, and it is undergoing a transition to become a more open economic and political system and experiencing dynamic reforms that can have a tremendous impact on all of CA.

Chapter 6 then focuses on the way the East Asian states' influences are shaped and received in CA using the case of the Chinese road maps of cooperation with Uzbekistan after the death of its dictatorial leader Islam Karimov and its re-opening to the world. By focusing on the impact of Chinese engagement in Uzbekistan, this chapter promotes an understanding of the motivations of smaller Central Asian states such as Uzbekistan in strategically engaging China. Although China's Belt and Road Initiative has received wide coverage, few details of its impacts have been analyzed. This chapter aims to fill this gap by outlining the latest project details of Chinese engagement and their impact in this region.

Chapter 7 outlines the issues and challenges related to the Japanese policy engagement in CA in an effort to distill lessons for improving its efficiency. As is narrated in several of the preceding chapters, one of the first countries in East Asia that applied the notion of Silk Road to its diplomatic initiatives in CA was Japan. The Japanese Silk Road Diplomacy, launched in 1997 by PM Hashimoto Ryutaro, following the Obuchi Mission of the same year, has become one of the first international diplomatic initiatives appealing to the connectivity and revival of the Silk Road. This has been followed by the successor initiatives by PM Koizumi Junichiro, who first dispatched a "Silk Road Energy Mission" in July of 2002 and launched the CAJ Japanese regional building initiative in August 2004. He also visited CA in 2006. PM Abe Shinzo visited all five CA states in 2015. These initiatives demonstrate that CA is Japan's latest "frontier" in Asia where Japanese presence can be further expanded. For CA governments, Japanese involvement in CA represents an attempt to balance Russian and Chinese engagements, while having access to the much needed technologies and knowledge needed to upgrade the industrial structure of their economies.

And finally, Chapter 8 provides a detailed analysis of the South Korean involvement in Uzbekistan, giving an outline of ambitious plans for developing a strategic partnership between the two countries. As is indicated in the following chapters, South Korea may not be the largest economic partner of Uzbekistan, in terms of volumes of trade and investment. But its structure of cooperation is certainly one of the most diverse, having spread through a number of sectors of the economy, making its impact on Uzbekistan's economy very significant. This chapter also details the competitive advantages of South Korea with respect to the Chinese and the Japanese initiatives.

Arguments on the Japanese, Chinese and South Korean engagements in CA

In addition to the arguments of each chapter elaborated in this study, the following general arguments are applicable to all the chapters of this study from both theoretical and empirical perspectives.

First, in terms of theory, this study argues against positivist (realist and liberalist) interpretations of the engagement strategies of China, South Korea and Japan by suggesting that the Silk Road and its "others" are socially constructed foreign policy agendas that arose from attempts by these states to react to changing foreign and internal environments. These foreign policies are also an attempt to socially shape the image of "others" to win the hearts and minds of targeted states by demonstrating how the (Chinese, South Korean and Japanese) "self" varies and is much more advantageous than alternative "others". In contrast to the positivist argument that such social constructs are formed against the common "enemy" or for a particular political/economic purpose (be it the expansion of Chinese corporate interests or Russian neo-colonialism), this study argues that these schemes do not have predetermined goals and aims but rather shape themselves according to changing perceptions and discursive categories of "self", "threats" and "opportunities". In addition, CA actors influence how these constructs are shaped by proposing their own visions of what they should be, thus sometimes challenging the visions of greater powers.

Second, when considering international engagement and cooperation schemes, scholars argue that they either develop from ideational commonality or from a practical functionalist logic of benefits and spill-over. In the case of the Silk Road imageries, this study demonstrates that the engagements envisaged by China, Japan and South Korea are socially constructed schemes that are related to shared norms, values and patterns of identities, in line with constructivist logic. This study further argues that functionality, pragmatism and the problem-solving focus within these engagements as well as the functionalist spill-over effect do not contradict constructivist explanations but can be regarded as generating norms and common values. States engaging in cooperation engage in certain interactions that can lead to trust, mutual understanding and spill-over into other fields and areas based on the newly emerged sense of common belonging generated from successful problem-solving experiences.

Third, the Silk Road rhetoric is not a defined foreign policy construct; rather, it is an engagement strategy that is easy for a target audience of states to comprehend and thus accept. This rhetoric, partly adopted for its resemblance to the historical trade routes, has a political meaning that is constantly contested by the various actors that choose to adopt it.

Fourth, one of the most interesting features of China's OBOR initiative (and its Japanese and South Korean alternatives) is that it features both decolonizing and neo-colonizing potential with respect to CA states. At the same time, Japan and South Korea emphasize only the decolonizing potential of their engagement with the region, thus using the absence of neo-colonizing features (similar to the Chinese ones) from their schemes to obtain a competitive advantage.

Notes

1 For details, see Timur Dadabaev, "Japan's Search for Its Central Asian Policy: between Idealism and Pragmatism", *Asian Survey* 53 (2013): 506–532, doi:10.1525/as.2013.53.3.506.
2 For an analysis of Silk Road discourse construction, see Timur Dadabaev, "'Silk Road' as Foreign Policy Discourse: The Construction of Chinese, Japanese and Korean Engagement Strategies in Central Asia", *Journal of Eurasian Studies* 9, no. 1 (2018): 30–41, doi:10.1016/j.euras.2017.12.003.
3 For details see Dadabaev, "'Silk Road' as Foreign Policy Discourse".
4 R Hashimoto, Address to the Japan Association of Corporate Executives. Tokyo, July 24, 1997, www. Japan.kantei.go.jp/0731douyuukai.html.
5 Zaidan Hojin Kokusai Koryu Senta [Japan Center for International Exchange], ed., *Roshia Chuo Ajia taiwa misshon hokoku: Yurasia gaiko he no josho* [Report of the Mission for Dialogue with Russia and Central Asia: Introduction toward Eurasian Diplomacy] (Tokyo: Roshia Chuo Ajia taiwa misshon, 1998).
6 Timur Dadabaev, "Discourses of Rivalry or Rivalry of Discourses: Discursive Strategies of China and Japan in Central Asia", *The Pacific Review* (2018, print forthcoming 2019), doi:10.1080/09512748.2018.1539026.
7 Timur Dadabaev, "Engagement and Contestation: The Entangled Imagery of the Silk Road", *Cambridge Journal of Eurasian Studies* 2018, no. 2, doi:10.22261/cjes.q4giv6.
8 For details, see, Timur Dadabaev, "Uzbekistan as Central Asian Game Changer? Uzbekistan's Foreign Policy Construction in the Post-Karimov Era", *Asian Journal of Comparative Politics* 4, no. 2 (2018, print forthcoming, 2019): 162–175, doi:10.1177/2057891118775289.
9 For discourse methodology, see David Machin and Andrea Meyer, *How to Do Critical Discourse Analysis. A Multimodal Approach* (London: Sage, 2012); Isabella Fairclough and Norman Fairclough, *Political Discourse Analysis: A Method for Advanced Students* (London: Routledge, 2013).
10 Timur Dadabaev, "Japan Attempts to Crack the Central Asian Frontier", *AsiaGlobal Online*, August 30, 2018, pp. 1–4.
11 See for instance, Ksenia Kirkham, "The Formation of the Eurasian Economic Union: How Successful Is the Russian Regional Hegemony?", *Journal of Eurasian Studies* 7, no. 2 (July 2016): 111–128; Carla P Freeman, "New Strategies for an Old Rivalry? China–Russia Relations in Central Asia after the Energy Boom", *The Pacific Review* 31, no. 5 (2018): 635–654, doi:10.1080/09512748.2017.1398775.
12 See for instance Anastassia Obydenkova, "Comparative Regionalism: Eurasian Cooperation and European Integration. The Case for Neofunctionalism?", *Journal of Eurasian Studies* 2, no. 2 (July 2011): 87–102. Also see, Rashid Alimov, "The Shanghai Cooperation Organisation: Its Role and Place in the Development of Eurasia", *Journal of Eurasian Studies* 9, no. 2 (July 2018): 114–124.
13 See for instance Yongquan Li, "The Greater Eurasian Partnership and the Belt and Road Initiative: Can the Two Be Linked?", *Journal of Eurasian Studies* 9, no. 2 (July 2018): 94–99.

14 For the purpose of simplicity, I refer to South Korea and Korea interchangeably in this book. In no parts of this book does Korea imply North Korea. For individual case studies, see Matteo Fumagalli, "Growing Inter-Asian Connections: Links, Rivalries, and Challenges in South Korean–Central Asian Relations", *Journal of Eurasian Studies* 7, no. 1 (2016): 39–48, doi:10.1016/j.euras.2015.10.004; Timur Dadabaev, "Japan's ODA Assistance Scheme and Central Asian Engagement", *Journal of Eurasian Studies* 7, no. 1 (January 2016): 24–38.

15 See Timur Dadabaev, "Chinese and Japanese Foreign Policies towards Central Asia from a Comparative Perspective", *The Pacific Review* 27, no. 1 (2014): 123–145, doi: 10.1080/09512748.2013.870223. Also see Timur Dadabaev, "Japanese and Chinese Infrastructure Development Strategies in Central Asia", *Japanese Journal of Political Science* 19, no. 3 (2018): 542–561, doi:10.1017/S1468109918000178.

16 See Dadabaev, "'Silk Road' as Foreign Policy Discourse"; Dadabaev, "Engagement and Contestation".

17 On the evolution and deconstruction of the "win–win" principle, see Rosita Dellios, "Silk Roads of the Twenty-First Century: The Cultural Dimension", *Asia & the Pacific Policy Studies* 4, no. 2 (2017): 225–236, doi:10.1002/app. 5.172.

18 Kyrgyzstan expressed concern about the route of the Kashgar–Osh–Andizhan railroad, which, from Kyrgyzstan's perspective, benefited China more than Kyrgyzstan, while Kyrgyzstan remained on the transportation route but saw no effects with respect to diversifying its economy and industrial structure.

19 For details see Timur Dadabaev, "Shanghai Cooperation Organization (SCO) Regional Identity Formation from the Perspective of the Central Asia States", *Journal of Contemporary China* 23, no. 85 (2014): 102–118, doi:10.1080/10670564.2013.809982.

20 Declaration on the Fifth Anniversary of the Shanghai Cooperation Organization, June 15, 2006, Shanghai, China, www.china.org.cn/english/features/meeting/171589.htm.

21 Shanghai Cooperation Organization (SCO), Charter of the Shanghai Cooperation Organization, June 2002, en.sco-russia.ru/load/1013181846, downloaded January 25, 2018.

22 M Oresman, "Repaving the Silk Road: China's Emergence in Central Asia", in *China and the Developing World: Beijing's Strategy for the Twenty-First Century*, ed. J Eisenman, E Heginbotham and D Mitchell (New York: M. E. Sharpe, 2007), pp. 60–84.

23 Timur Dadabaev, "Shanghai Cooperation Organization (SCO) Regional Identity Formation".

24 For details, see Office of the Leading Group for the Belt and Road Initiative, *Building the Belt and Road: Concept, Practice and China's Contribution* (Beijing: Foreign Languages Press, May 2017), especially pp. 11–17, https://eng.yidaiyilu.gov.cn/wcm. files/upload/CMSydylyw/201705/201705110537027.pdf, accessed March 12, 2018.

25 See National Development and Reform Commission, Ministry of Foreign Affairs and Ministry of Commerce of People's Republic of China, "Vision and Actions on Jointly Building Silk Road Economic Belt and 21st-Century Maritime Silk Road", March 30, 2015.

26 As an example of such new engagement, see Ministry of Foreign Affairs of the People's Republic of China, "Joint Declaration on New Stage of Comprehensive Strategic Partnership Between the People's Republic of China and Republic of Kazakhstan", August 31, 2015, www.fmprc.gov.cn/mfa_eng/wjdt_665385/2649_ 665393/t1293114.shtml, last seen on January 25, 2018.

27 See, for instance, "Posol: proekt 'Odin poyas, odin put' – eto ne ekspansiya Kitaya" [Ambassador: "One Belt, One Road" – is not expansion of China], *Podrobno.uz*, May 29, 2017, www.podrobno.uz, last accessed on May 29, 2017.

28 M Auezov "Ex-Ambassador of Kazakhstan to China Concerned over China's Classified Documents", *Tengri News*, 2015, https://en.tengrinews.kz/politics_sub/Ex-Ambassador-of-Kazakhstan-to-China-concerned-over-Chinas-21645/, last accessed

on January 15, 2016. Also see M Auezov, "Kazakhstan Must Stop Wavering between Russia and China to Pursue Central Asian Consolidation", *Interfax*, January 29, 2013.

29 R Khodzhiev, "V svoyom dome ne khoziaeva: Kitaiskaia ekspasiya v Tazhikistan", *Centrasia.ru*, November 1, 2011, www.centrasia.ru/news.php?st=1320094800, last accessed on November 1, 2011).

30 "Stroitel'stvo zh/d Chuy–Ferghana vygodnaya al'ternativa doroge Kitai–Kyrgyzstan–Uzbekistan: expert" [Construction of the railroad Chuy–Ferghana is beneficial alternative to the road of China–Kyrgyzstan–Uzbekistan: expert], *KyrTag (Kyrgyz Telegraph Agency)*, April 20, 2012, www.kyrtag.kg/?q=ru/news/19472, accessed on April 20, 2012.

31 For details see, Timur Dadabaev, *Japan in Central Asia: Strategies, Initiatives, and Neighboring Powers* (New York: Palgrave Macmillan, 2016).

32 "Chuou Ajia hatten no kokusaiteki jyoken to Nihon" [Development of Central Asia: International conditions and Japan], *Gaiko* 34 (2015): 21–34, especially p. 21.

33 Joint statement of Tajik President and Japanese PM, 2015, Ministry of Foreign Affairs, Japan.

34 Joint statements of Japanese PM and CA Presidents, 2015, Ministry of Foreign Affairs, Japan.

35 M Rakhimov, "Central Asia and Japan: Bilateral and Multilateral Relations", *Journal of Eurasian Studies* 5, no. 1 (2014): 77–87, doi:10.1016/j.euras.2013.09.002.

36 Tony Tai-Ting Liu, "Undercurrents in the Silk Road: An Analysis of Sino-Japanese Strategic Competition in Central Asia", *Japanese Studies* 8 (March 2016): 157–171.

37 For details, see interview with FM Fumio Kishida, "Shinrai to sougo kanshin no kiban no ueni" [Building on the foundations of trust and mutual interest], *Neutral Turkmenistan*, May 9, 2017, p. 3, in Japanese and Russian at www.mofa.go.jp/mofaj/p_pd/ip/page4_002989.html.

38 See, for instance, the roundtable discussion "Chuou Ajia hatten no kokusaiteki jyoken to Nihon", especially pp. 29–33.

39 See "Chuou Ajia hatten no kokusaiteki jyoken to Nihon", especially p. 31.

40 "Shyusho 'Ittai ichirou ni kyoryoku' Hatsuno hyoumei" [PM first announcement of preparedness for cooperation along "OBOR/BRI"], *Yomiuri Shimbun*, June 5, 2017.

41 "Chugoku no 'Ittai Ichiro' Kyoryoku ni Seimeisei, Kouseiseinadoga 'jyouken'" [Transparency and legitimacy are the conditions for support of Chinese OBOR], *Sankei Shimbun*, June 5, 2017, www.sankei.com, last seen on June 8, 2017.

42 See "Foreign Ministry Spokesperson Hua Chunying's Regular Press Conference on June 6, 2017", www.fmprc.gov.cn/mfa_eng/xwfw_665399/s2510_665401/t1468248.shtml.

2 Evolution of the "Silk Road" into a foreign policy discourse

The collapse of the Soviet Union has resulted in the emergence of a number of new regions. The southern periphery of the Soviet Union, consisting of five stans (Kazakhstan, Kyrgyzstan, Tajikistan, Turkmenistan and Uzbekistan), is one of these newly emerging regions. In the Soviet era, this region was divided into Middle Asia and Kazakhstan; in the post-Soviet era, it has been largely referred to as Central Asia (CA). In geopolitical terms, however, this area has often been referred to as the Silk Road region. Many states have used the rhetoric of reviving the Silk Road to imply closer engagement with the CA region and its eventual integration into a network of economic ties. For instance, as has been mentioned in the introductory chapter, in its first initiative in the Eurasian region, Japan launched its Silk Road Diplomacy as early as 1997 under Prime Minister (PM) Hashimoto's administration.[1] In 1999, the US proposed the Silk Road Strategy Act to expand the US presence in this region and to sustain Russian dominance.[2] South Korea subsequently launched a number of similar strategies through 2009–2013 under the "Silk Road" umbrella, to connect South Korea through Russia, China and CA railroad networks to energy and other resources in Eurasia. Finally, the Chinese official discourse has evolved from an emphasis on reviving the Silk Road through constructing Eurasian Land Bridges in the early 1990s to establishing the "One Belt, One Road" (OBOR, and Belt and Road Initiative, BRI) initiative in 2013 to revive the Silk Road region by linking China to other markets through the various infrastructure networks of CA.

The content and the nature of these Silk Road strategies changed with time and the international environment. The concept carried different meanings depending on the country that chose the Silk Road as a brand for its foreign policy engagement with the CA region and evolved depending on the tasks each of the countries employing it sought to accomplish through their "Silk Roads". In this sense, the notion of the Silk Road has changed from a static concept of a historical trade route into a product of social construction upon which various states have built their relations within the CA region and beyond. Thus, the Silk Road as a term has come to represent the various CA engagement strategies of a number of powerful states – strategies that are constantly shaped, imagined and socially constructed. In this sense, the Silk Road is not a foreign policy doctrine but rather a discursive strategy of engagement that largely exists in the realm of narration. This narration is also a matter of social construction that is subject to change depending on the international environment of the country (China, Japan, Korea, etc.) that produces such narratives, the alternative narratives that compete for wider international acceptance and the country's vision of "self" and the "other" in the international context.

This chapter thus focuses on the manifestations of such Silk Road narratives in Chinese foreign policy and their Japanese and South Korean counterparts and raises the following questions: How do the "Silk Road" narratives in the Chinese, Japanese and Korean engagements in CA evolve? What is the process of their social construction? The primary objective of this chapter is to address these questions and to stimulate debate on the use of the Silk Road narrative among both academics and policy makers.

Thus, by seeking an answer to these questions, this chapter aims to demonstrate the evolution of the Silk Road narrative historically, geographically and in terms of the existential maturation of this foreign policy concept. Additionally, this chapter aims to highlight the similarities and differences in the construction of the Silk Road narrative by expert (policy and academic) communities to demonstrate the mutual relevance of the Silk Road to the states seeking relations with the CA region.

More specifically, this chapter aims to demonstrate that the discursive practice of using the term "Silk Road" has different connotations and meanings in different (Chinese, Japanese and South Korean) settings. Such differences also suggest that although countries may brand their endeavors using the concept of the "Silk Road", this does not necessarily mean that they have similar goals and objectives and that their initiatives are therefore characterized by exclusivity and rivalry. On the contrary, as seen in different parts of this chapter, the varying meanings of the "Silk Road" framing leave space for complementarity in these initiatives.

Internalizing the post-Soviet vacuum in Chinese, Japanese and Korean engagements in CA

China, Japan and South Korea faced similar uncertainties in their relations with post-Soviet states following the collapse of the Soviet Union and the eventual independence of CA republics. Lack of a clear understanding of the foreign policies of the newly independent states complicated the task of formulating clearly defined interests in the region. In the aftermath of the Cold War, both China and Japan also confronted changes in their own economies and international standing and in the development of their international affairs. Therefore, the interests and foreign policies that they developed toward this region were largely socially constructed through interactions with these new states, as seen in the dynamics of the constantly changing agendas of the Chinese (Silk Road to Shanghai Five toward the Shanghai Cooperation Organization (SCO) and OBOR/BRI), Japanese (Eurasian/Silk Road Diplomacy toward CA plus Japan and PM Abe's CA policy) and South Korean (Roh Moo-hyun's Comprehensive Central Asian

Initiatives of 2006 to Lee Myung-bak's New Asia Initiative of 2009 and Park Geun-hye's Eurasia of 2013) initiatives.

The collapse of the Soviet Union was an unprecedented and unpredictable event that left many inside and outside the USSR uncertain about how to conceptualize approaches toward this region. This uncertainty was felt in both China and Japan immediately after the fall of the USSR. It is not a coincidence that the first proper studies of the CA region were not published in China until the mid-1990s. Attempts to shape the understanding of the Chinese expert community appeared to intensify beginning in approximately 1995 with the publication of *Zhōng yà Yan'zyu* [Research into CA], *Zhungo sin duli de zhōng yà godzya guansi* [Relations of China with newly independent states], *Zhōng yà wǔ guó gàikuàng* [The general situation of five CA states], *Zhōngguó yui Zhōng yà* [China and Central Asia] and others.[3] These studies offer initial insights into the domestic and foreign policies of CA states and attempt to integrate them into the Chinese foreign policy agenda. However, as is stated in these and similar studies, although relations with CA states were of high priority, they were not considered of crucial importance for China at the time.[4]

The primary objective of Chinese foreign policy at the early stage of CA states' independence was not to construct a cohesive and long-term Chinese dominance strategy in this region. Rather, the Chinese expert community and general governmental policy were attempting to pre-empt the negative influence of CA states on Chinese domestic and foreign policies. In addition, in the context of the post-1989 intensification of anti-China sentiment in the West following the Tiananmen Square protests, China was attempting to build a coalition of China-friendly countries.[5] During this period, China's internally oriented policies also showed the first signs of shifting, with greater attention being devoted to the intensification of Chinese foreign policy in the aftermath of the collapse of the USSR. Such intensification was both "defensive" in its intention to counter the negative impact of the events of 1989 on the international image of China and "offensive" as China sought new opportunities in the aftermath of the collapse of the USSR.[6]

The new environment in the CA region that China had to adapt to required new patterns of interaction, with negative and positive consequences. In terms of the negative consequences, the following were mentioned by various Chinese authors as requiring careful attention.

First, China had previously had to consider the interests of only one state, namely, the USSR. With the collapse of the USSR, Chinese policy makers realized that they had to manage relations with a multiplicity of states, which required greater flexibility with respect to this region.[7] Interestingly, many conclude that, in the early years of CA independence, China did not aim to benefit from interactions with CA states and was not engaged in rivalry with any states in this region. Such rhetoric is similar to the rhetoric of the Japanese government described below.

Second, the Chinese agenda with respect to CA in the early 1990s did not involve economic expansion into this region but focused more on resolving territorial disputes and preventing support for separatism. In addition, Chinese policy makers emphasized that China needed to consolidate regional support for the "one China policy" and prevent CA states from recognizing Taiwan.[8]

Third, pan-Turkic and pan-Islamic calls in some CA states were received as alarming signs in China due to their potential expansion into Chinese territories.[9]

Fourth, though post-Soviet CA republics were endowed with natural resources, the potential development of these resources was complicated by the differences in the states' approaches and lack of common vision, which hindered the regional integration of these states. These differences, according to the Chinese expert community, were considered a weak point that could be exploited by groups with evil intentions. In addition, Chinese policy makers regarded regional rivalry between CA states as a factor that further weakened their prospects for cooperation and development.[10]

Fifth, China regarded initiatives by other states in the CA region in the late 1990s as attempts by larger states to exploit this region (in the case of the US, as a containment strategy against Russia or China,[11] and in the case of Russia, to consider CA as its "backyard"[12]) which China was not prepared to accept and tolerate. This last point does not necessarily imply that Chinese discourse on CA has been shaped by a decolonizing motivation. Rather, it implies that China was willing to oppose attempts by other countries to consider the newly created vacuum in CA as an opportunity to expand their dominance. This principle is not new to Chinese foreign policy, but in the CA context, it has been warmly received by CA counterparts who also feared that their relatively small countries would eventually become victims of great-power politics.[13]

The launch of the SCO was largely a response to these negative consequences and introduced a mechanism to address all of these Chinese concerns.[14] While the SCO's objectives and the Shanghai spirit in the decision making of this organization implied security for all participating states, the Chinese rhetoric about this organization emphasized its importance for Chinese interests, while regional security was used as a tool primarily to secure Chinese interests.[15] In this sense, the creation of this organization was pragmatic and utilitarian rather than focused on common identity formation.[16]

In contrast with the negative consequences associated with the collapse of the Soviet Union, as already mentioned, there were several opportunities for Chinese participation that Chinese policy makers considered to be positive side effects of the collapse of the Soviet Union.

First, CA states received a chance to decolonize themselves and escape the dominance of Russia. Thus, Russian pressure in the areas bordering China diminished as China's border with Russia shrank and "buffer states" were created between Russia and China.

Second, the independence of CA states offered new opportunities for China to develop alternative transportation infrastructure that would allow China to access European markets using the territory of the newly independent states rather than exclusively relying on Russian railroads. Although Chinese experts have not considered Russia to be an imminent threat to Chinese corporate interests, they considered CA alternative railroads to be a safeguard against potential Russian pressure should relations between China and Russia worsen in the future. In recent years, Chinese policy makers have adhered to the policy of creating as many alternatives as possible, as demonstrated by the construction of a railroad between Uzbekistan and China, the intent of which is to balance reliance on Russian and Kazakh railroads by constructing an alternative one running through Kyrgyz territory.

Third, the CA region can play a significant role in China's ability to secure energy supplies.[17] According to the Chinese narrative, this needs to be done by promoting the active participation of Chinese corporations in oil production, thus ensuring China's energy security.

Fourth, the emergence of a number of smaller states between Russia and China allowed the Chinese government to reduce the number of troops stationed along the border with the former USSR as a result of the decreased threat posed by Russia and the increased security provided by the buffer states positioned between Russia and China.

Similarly, Japan was also motivated to redefine its place in the changing international environment with the disappearance of the Soviet Union, the emergence of newly independent states and the need to redefine its relations with Russia and China.[18] Thus, efforts to establish a Japanese engagement strategy in the CA region were launched in the form of Eurasian (Silk Road) Diplomacy (1997–2004) by PM Hashimoto Ryutaro, who was a believer in interdependence, terming it "mutual necessity".[19] He hoped to bring the nations of the former Soviet Union into a network of interdependence by establishing a largely economic and, to a degree, political Japanese presence in Eurasia and by facilitating Japanese participation in resource exploration. Although Japan achieved its aim of providing much needed economic and humanitarian assistance to the newly independent states, other goals of the initiative were hardly achieved. The failures of these early Japanese engagements were rooted in the following reasons.

First, despite historical references to the Silk Road connections, Japanese foreign policy poorly defined CA, as its policy in Asia often excessively focused on Association of Southeast Asian Nations (ASEAN) countries.[20] Even in his Eurasian (Silk Road) Diplomacy speech, PM Hashimoto stated that "the primary objective of Japan's foreign policy was to maintain the peace and prosperity in the Asia-Pacific region", while the Silk Road region was defined as an ambiguous new frontier.[21] In this sense, the task of "defining" the importance and place of CA for Japan has been and remains one of Japan's greatest challenges due to its relative distance from the CA region, which makes it more difficult for Japanese policy makers to "frame" this region's importance for Japan in practical terms. Some scholars suggest that this lack of interest is an advantage for the Japanese foreign policy approach in this region, as Japan could claim to be motivated by interest in developing the region rather than by possible benefits for Japan.[22] As recognized by several diplomats who established Japanese embassies in the region, most of the initiatives were achieved spontaneously through the enthusiasm and commitment of ambassadors and officials from the

Ministry of Foreign Affairs (MOFA), Ministry of Finance and other related agencies. In 1997, PM Hashimoto referred to the ASEAN Regional Forum and Asia-Pacific Economic Cooperation in his Eurasian speech as the main platforms "to stage Japan's basic foreign policy". His Eurasian Diplomacy reflected such limitations and called for a needed "push to enlarge the horizon of our foreign policy" beyond the Asia-Pacific, building new ties with Russia, China, the Central Asian republics and the Caucasus.[23] However, even the "CA plus Japan" dialogue scheme of 2004 was modeled on Japanese engagement with ASEAN countries by borrowing the "ASEAN Plus 3" model and applying it to CA.[24] Some at the MOFA and in the expert community now suggest that this scheme can be used to create an ASEAN-like organization in the CA region.[25]

In terms of institution building, in contrast to China and Russia, Japan advocates the notion of "open regional cooperation" in CA. All the Japanese prime ministers who championed CA engagement – including PM Hashimoto, with his Eurasian Diplomacy; PM Koizumi, with his CA plus Japan initiative; and PM Abe, with his CA policy – referred to inclusive regional institutions (which do not exclude Russian and Chinese participation) that do not limit other countries' participation. There are several reasons for such an approach. Japan traditionally supports fair and unrestricted regional cooperation in other parts of the world, as inclusive regional institutions offer members more flexibility than closed regional institutions. PM Hashimoto's initiative emphasized the importance of Asian countries, with PM Hashimoto going so far as to say that "world diplomacy has shifted from axis of the Atlantic Ocean and Europe ... to axis spanning the Eurasian landmass encompassing many nations, small and large".[26] The official Japanese discourse on the CA plus Japan initiative created under PM Koizumi's watch also emphasizes that, by advocating open regionalism, Japan does not intend to curtail Chinese or Russian interests in this region.

Japan aims to use its distance from this region to gain a competitive advantage (when compared to other countries like China and Russia): it attempts to position itself as a neutral mediator for CA states by suggesting that its distant geographic location prevents it from dominating and exploiting CA states. In this sense, the official Japanese discourse states that this scheme was designed to encourage CA states to seek alliances with each other while Japan would provide the technical and financial assistance needed to support such alliances. This objective of the CA plus Japan initiative is rooted in the legacy of PM Hashimoto's Eurasian (Silk Road) Diplomacy. According to this initiative it, it is "important for Japan to promote regional cooperation aiming to create a transport, telecommunications and energy supply system and for Japan to cooperate in developing energy resources in that region".[27]

Seeing common values as the basis for cooperation, Japanese foreign policy favors the notion of partnerships rooted in universal values (such as democracy, a market economy, the safeguarding of human rights and the rule of law). In this sense, Japan did not develop norms of behavior, similar to the Shanghai spirit. However, as many in the Ministry of Foreign Affairs of Japan believe, institutionalization of the global values mentioned above has been an over-inflated expectation, and an understanding of the importance of cooperation among CA states is lacking.[28] Thus, while Japan hopes for the institutionalization of such universal values, these values are not used as a precondition for its aid programs. Japanese officials often display sympathy toward CA states' difficulties in adopting universal values, claiming that Japan was also a recipient of economic aid before it attained the status of an economic superpower.[29] In a sense, such duality of approach became a certain common norm of behavior between CA states and Japan: Japan does not completely abandon its commitment to universal values but does not use it as precondition for cooperation, offering CA states an opportunity to adjust and build their domestic conditions for implementation of these universal values.

Retrospectively, PM Hashimoto's Eurasian (Silk Road) Diplomacy initiative of 1997 was followed by the Central Asia plus Japan initiative implemented in 2004 under the PM Koizumi administration to expand Japanese participation in this region and establish an organization through which Japan and other CA counterparts could discuss issues vital to CA regional development. The establishment of this initiative was followed by PM Koizumi's visit to Uzbekistan and Kazakhstan in 2006. To further integrate this region into Japanese foreign policy, FM Aso Taro proposed the concept of "Central Asia as a Corridor of Peace and Stability" in 2006, which involved incorporating this region into larger Japanese initiatives in the Middle East, to which CA is also conceptually connected.

Since the 2015 visit of PM Abe, a more goal-oriented, practical approach to cooperation with CA focused on functionality and the practical outputs of such cooperation has been prioritized over the value-based approach. This shift may be due to a realization by Japanese leadership that, for CA, the process of democratization is a longer-term objective, and in the meantime, the economic opportunities of cooperation need to be taken.

South Korea has the same limitations as Japan, with its distant location and lack of transportation infrastructure to and from CA markets. However, South Korean "Silk Road" rhetoric is somewhat more practical than that of Japan. South Korean corporate economic interests were present and more active in the region beginning in the early 1990s, with Daewoo building a major car manufacturing plant in Uzbekistan; in addition, a large number of assembly and manufacturing facilities were built under the Samsung and LG brands in Uzbekistan and Kazakhstan to locally produce electronics parts. In the early 1990s, South Korea was among the leading investors in this region. In addition to building the car manufacturing plant in Andijan, South Korea has also been involved in setting up a number of enterprises that have built and sustained the Uzbek and Kazakh economies. Among these were the Daewoo Unitel (communications company) and Kabool Textiles (cotton processing and textile production company) plants in Uzbekistan and the Samsung assembly plants in Kazakhstan.[30]

However, conceptually, engagement with CA received a boost with the President Roh Moo-hyun administration of 2003–2008. This period featured two visits to Kazakhstan and Uzbekistan in 2004 and 2005, respectively, followed by the international conference launched by the government of South Korea in December 2005. As a result of that meeting which included participants from CA countries (including governmental representatives and various public organizations), the South Korean government "Comprehensive Central Asia Initiative" was formulated, and adopted in 2006. The goal of the program was to establish a proper standing for South Korean interests in Eurasia and, in practical terms, to secure the long-term energy resources supply.

In 2006, Uzbek president Karimov and his Korean counterpart signed the Joint Declaration on Strategic Partnership, providing a new framework for ROK (Republic of Korea) investments in Uzbekistan. This paved the way for extraction agreements in 2006 between Korea National Oil Corporation (KNOC) and Uzbekneftegaz granting KNOC exclusive exploration rights over Chust-Pap and Namangan-Terachi.

Such intensification of contacts between South Korea and CA states resulted in the establishment of the Korea–Central Asia Forum, the first occurrence of which was held in Seoul in November 2007. The goal of this forum is to create a platform for enhanced exchange of ideas between government and private-sector participants regarding possible ways to enhance cooperation between South Korea and CA states.[31] Up until 2017, the forum was held annually. However, in 2017, as if reflecting the need for further intensification, The Korea–Central Asia Cooperation Forum Secretariat was established in Seoul as a joint permanent body.[32] Although the branding and the function of this forum is somewhat similar to the Japanese one, the implementation and structure of the Japanese forum includes several tracks of dialogues (senior officials' meetings, dialogues of intellectuals, etc.) without having a permanent body. In a sense, the Japanese platform aims to shape otherwise chaotic tracks of dialogues while the Korean forum aims to organize a permanent institution for cooperation. As of now, the efficiency of both bodies is unclear as no clearly defined outcomes have been produced by either.

Prime Minister Han Seung-soo's visit to Central Asia in May 2008 resulted in South Korea signing a contract to purchase from Uzbekistan 2,600 tons (worth around US$400 million) of uranium between 2010 and 2016 which equaled an amount roughly 9 percent of South Korea's annual uranium consumption. A similar agreement has also been concluded between South Korea and Kazakhstan. In particular, KNOC agreed on a deal worth US$85 million to obtain a 27 percent stake in Kazakhstan's Zhambyl oilfield, located in the Caspian Sea. According to this agreement, the South Korean corporation agreed to jointly explore that oilfield together with KazMunaiGaz. In addition to the interest in energy resources, and as well as a humanitarian assistance plan which included US$120 million economic aid to Uzbekistan and additional aid for improvement of medical facilities, there were few other factors which influenced South Korean motivations in CA. As mentioned above, South Korea has also been proactive in transportation infrastructure construction, with the Hanjin Group establishing and developing a Navoi logistics hub.[33]

After a visit in 2009, President Lee Myung-bak again visited CA in 2011 and participated in opening several enterprises with South Korean capital in Uzbekistan and Kazakhstan. Thus, geographically, the South Korean Silk Road in the CA region largely refers to these two countries, while other countries are of lesser interest.[34] Similar to the Chinese and Japanese concepts, the Korean Silk Road strategy also evolved in response to the growing opportunities for Korean enterprises in CA focusing on energy resources and logistics as well as manufacturing. In this sense, the practical application of South Korean initiatives somewhat resembles the Chinese and the Japanese ones in an attempt to pragmatically use emerging new opportunities. However, the Chinese, Japanese and Korean Silk Road projects are somewhat different in scale of coverage, agenda and primary motivation. In this sense, Korean engagement under the umbrella of the Silk Road largely represents a branding strategy for economic engagement rather than being an ideological driver, as seen in the Chinese and Japanese patterns of establishing their presence in CA.

As a side effect of such successful cooperation, Uzbek president Karimov appointed a Korean national to serve in his government charged with the task of building the e-governance system, which signified the degree of trust toward South Korea. Eventually, e-governance was largely considered to be a success story which facilitated the appointment of another Korean national into the government under the new President Mirziyoyev's administration. In addition, notions of smart cities and urban management as well as systems used in South Korea have been accepted as the modes for application in Uzbekistan. In this sense, the South Korean influence has been felt not only in the area of economic cooperation but also in political and societal modernization, as detailed in Chapter 8 of this volume.

Chinese, Japanese and Korean politics of the "Silk Road"

As stated above, the notion of the Silk Road is contested; there is no consensus on its meaning. As already observed, the Silk Road frequently represents a preferred framing for engaging CA states and other FSU (former Soviet Union) countries, and it has become a contested concept in terms of both its constituent principles of such engagements and the region that it covers.

The Chinese narrative regarding CA was initially a response to various challenges that China regarded as constraints on its own position in this region. Thus, the initial attempts at conceptualizing a new Chinese positioning in this region were made in response to such challenges. However, after 2000 and with the successes of the SCO, China sought to consolidate its security, economic and political agendas under the same umbrella. One of the most difficult tasks was to frame this new agenda in a way that would make it acceptable to CA states and Russia. For this task, the notion of the Silk Road and its historical connotations proved highly useful.

The Japanese Silk Road narrative and its modern re-emergence are connected to the values of common historical heritage that support cooperation between Japan and CA countries. PM Hashimoto, in his speech, emphasized that "Japan has deep-rooted nostalgia for this region stemming from the glory of the days of the Silk Road". He cited the common heritage of the Silk Road as "a solid foundation upon which to build firm relations with these (CA) countries as friendly states".[35]

For Japanese PM Hashimoto, the areas of interaction included, first, assisting these states in establishing affluent, prosperous domestic systems under a new political and economic structure. Additionally, by providing assistance, Japan aimed to help these states forge peaceful and stable external relations. Second, Japan aimed to utilize the great potential of these states to serve as bridges to create distribution routes within the Eurasian region.[36] In terms of particular directions, Japanese Silk Road Diplomacy channeled Japanese foreign policy toward CA states into three areas: political dialogue; economic cooperation, including cooperation for natural resource development; and cooperation in peace-building, nuclear non-proliferation, democratization and fostering stability. These areas and directions were largely inherited by the subsequent administrations. Interestingly, the prospective areas and principles of interaction within Japanese Eurasian (Silk Road) Diplomacy bear many similarities to the Chinese engagement principles, as explained below.

PM Hashimoto emphasized three main principles of Japanese engagement: establishing trust, establishing a mode of "mutual benefit" and "maintaining a long-term perspective" that would allow the outcomes to be inherited and further built upon by subsequent generations. These are also strikingly similar to the principles of Chinese engagement discussed below.

Under the second Abe administration, the same common heritage has been re-emphasized, with FM Kishida stating during his visit to Turkmenistan in May 2017 that

for many Japanese people, Central Asia is associated with the Silk Road, as it brought Buddhism, which is the basis of the Japanese culture, and enriched Japanese culture by introducing civilizational and cultural influences of the West; all were communicated through the Silk Road.[37]

The task of prioritizing areas of interaction with these CA states was another important issue for both China and Japan. Both countries attempted to build their relations with this region based on their historical connections with CA and on new Silk Road projects.

The Chinese frames of engagement in the CA region defined in the late 1990s resemble those outlined in Hashimoto's Eurasian Diplomacy platform. In particular, similarities can be found in defining areas of cooperation, in China's adherence to the principle of mutual benefit in economic relations and adherence to the market economy system, maintaining balance of imports and exports between China and CA states. In addition, the Chinese discourse on efficient cooperation outlines transportation and logistics as among the most important areas with high potential.[38]

Another similarity shared by the Japanese and Chinese frames of engagement can be found in the fact that both Japanese and Chinese official discourses assert that corporations should take the lead in facilitating economic projects. Both Japan and China emphasized the importance of facilitating transportation connectivity and access to CA markets through Chinese railroads and logistical hubs. However, as stated by Chinese scholars, many Chinese companies felt hesitant to enter CA markets without governmental guarantees and support because of the high risks related to these newly formed markets.[39] The Chinese government dealt with such concerns by establishing governmental support frameworks for the foreign expansion of Chinese corporations, as detailed in the section on the OBOR/BRI initiative below. However, for the Japanese corporate community, similar concerns existed and still exist as the main obstacles to wider Japanese participation in CA markets.

There are also similarities in terms of how China and Japan conceptualized the exclusivity of China–CA and Japan–CA relations, respectively. The Japanese concept of Eurasian Diplomacy in the Silk Road area emphasized region building based on the principle of open regionalism under the leadership of the Japanese government and including China and Russia. The Chinese discourse of the late 1990s portrays this region as one that has yet to define itself under the constantly changing circumstances, thus making it premature to define certain aspects or areas (be it security, trade or production) in which China could establish its own sphere of domination in this region.[40] Thus, Chinese policy and scholars depicted Chinese engagements in this area as hypothetically open to Russia and other states, including those within the SCO. Therefore, many in the Chinese academic community did not rule out that Japan could contribute and participate in this region together with China.[41]

In terms of the concept of providing stability and security to this region, again the notions outlined in Hashimoto's Eurasian speech and the contemporary discourse of Chinese academics and policy makers bear striking similarities. Both emphasize the importance of developing the economic structures of the CA region by encouraging the creation of production capacities, because both Japan and China frame this task as the most important for ensuring security. China appeared to be aware of the problem of cheap and low-quality Chinese products flooding CA, which the Chinese expert community believed would damage long-term Chinese interests and China's image in this region.

Finally, both Japan and China envisaged their own models of development for CA states. In the Hashimoto initiative, the PM of Japan emphasized the importance of facilitating the transition to a market economy using Japanese experience. Similarly, China also emphasized that the Chinese experience of modernizing the Socialist system might be useful for the CA states.[42] Thus, both countries claimed to possess a model of development that could serve as an example that CA states could adapt to their own conditions.

However, there are also differences in the Chinese and Japanese approaches. For instance, in the Japanese case, the government is the primary agent of Japanese penetration into CA, while other non-state actors are not delegated proper roles. In the Chinese case, policy makers and academicians emphasize the importance of not only government-to-government interaction but also region-to-region ties. Thus, areas of China with close geographic proximity to CA are expected to play significant, if not the most important, roles in this process.

The Korean usage of romanticized images of the Silk Road to emphasize common heritage and mentality is somewhat similar to the Japanese rhetoric. For instance, in 2014, President Park made reference to Ahal-Teke, a horse breed in Turkmenistan, and compared it to the history of economic development stating that

> Just as the Ahal-Teke became an excellent steed as a result of endless self-discipline, in order to survive in a rugged mountainous area, we too will be able to make the Eurasia Initiative a success if both the government and companies push forward by gathering power with the spirit of challenge and talent.[43]

However, the emphasis of her speech was more on energy and transportation, specifically the realization of the South Korean "Eurasia Initiative" of which the "Silk Road Express" is a consistent part. According to these plans, South Korea aims to connect the Trans-Korean (TKR) and Trans-Siberian (TSR) Railroads by linking Busan, China, Russia, Central Asia and Europe. This makes the usage of "Silk Road" branding by the South Korean government more compact and pragmatic than the Silk Road Diplomacy of Japan, as the South Korean government basically follows the lead of its corporations to secure wider access to CA markets.

In terms of evolution of South Korean initiatives, President Roh Moo-hyun's administration launched a "Comprehensive Central Asia Initiative" as early as 2006, serving as the basis for the "Energy Silk Road Diplomacy" established by President Lee Myung-bak.

President Lee, during his visit to CA in 2009, secured an agreement to build a petrochemical plant in Atyrau with a budget of US$4 billion, a US$4 billion contract to construct power generation plants in Balkhash and a US$4 billion deal to participate in development of a gas field in Uzbekistan.[44] In 2011, during another visit by President Lee, the construction of the Ustyurt gas chemical complex was launched. In addition, South Korea's Hanjin (parent company of Korean Air) received exclusive rights from Uzbekistan to develop its Navoi International Airport into an intercontinental logistics hub.

In the same year, the Ministry of Strategy and Finance announced that Kazakhstan and Uzbekistan would be receiving a package cooperation deal, while other smaller CA countries would benefit from a deal-to-deal approach.[45] While both China and Japan differentiate between smaller and larger countries in CA, the South Korean approach clearly distinguishes Uzbekistan as the most populous and energy-rich and Kazakhstan as the most economically viable and energy-endowed countries in the region. South Korea therefore approaches these countries with particular interest. Thus, geographically, the South Korean Silk Road in the CA region largely refers to these two countries, while other countries are of lesser interest.

President Park Geun-hye also supported the energy-related and transportation-focused South Korean engagement in CA by developing this initiative not only to include energy cooperation but also to expand the notion of the Silk Road to build a "Silk Road express" which is an idealized concept referring to the railroad to connect North and South Koreas, Russia, Central Asia and Europe.[46]

If this transportation network is realized, South Korea receives an additional transportation network for two pillars which are of crucial importance for the South Korean economy – namely, international trade and investment areas. As a result, this might lead to the diversification of its economic ties. In terms of principles of South Korean engagement, the Park administration's Eurasian Initiative envisaged several goals: first, enhancing partnerships with Eurasian states and integration with Eurasian states for development of new markets for South Korean companies; second, developing transportation linkages and trade networks to balance South Korean connections which are currently are focused on China and the US by developing the Eurasian direction. And third, facilitating a positive impact on North–South (Korea) relations by promoting economic growth through the networks and linkages mentioned above. Interestingly, similarly to the Chinese and the Japanese rhetoric, South Korean initiatives also emphasize the principle of "win–win" cooperation for all those involved.[47]

One additional factor for South Korean attempts to engage CA states, which drastically differs from both Japan and China, is the substantial ethnic Korean community mainly residing in Uzbekistan and Kazakhstan but also present in Kyrgyzstan and other states of the region.[48] As is well covered in the relevant literature, the Korean diaspora in CA is one of the largest in world. It emerged as a result of mass deportation policies in the late 1930s from the Soviet Far East to Central Asia. Under this policy, the Koreans were settled in CA. They were well received by CA populations and thus integrated into the structure of their societies, contributing to development of republics where they re-settled.

OBOR/BRI as the Chinese "self" and its Japanese/South Korean "others"

The idea of the revival of the Silk Road as a concept for the enhanced cooperation of the states along its historical route is not new. In 1992, immediately after the collapse of the Soviet Union, the re-emergence of the Silk Road was referenced by the foreign minister of China at that time, Qian Qichen, during his visit to Uzbekistan. In 1994, PM Li Pen also confirmed the importance of the Silk Road for relations with CA states and the potential for its re-emergence. However, the discussions regarding the intensification of contacts along the Silk Road were mostly focused on transport infrastructure development, as exemplified by the construction of the Second Eurasian Land Bridge (the first being the one connecting China to Europe through Russia), which was completed in September of 1990 and opened to full traffic in June 1992. The railroad has been used by many countries, including China and South Korea. The support for the Second Eurasian Land Bridge was overwhelming within the Chinese expert community (with some referring to it as a Golden Belt), as it offered the advantages of being shorter (and thus economically more beneficial), politically independent of Russian control (in the case of a hypothetical worsening of relations with Russia) and covering areas of smaller countries that cannot exert pressure on China.[49] The economic benefits of the second bridge have been obvious for China, as the railroad has brought in revenue from the transportation of goods through China. At the same time, it also benefited Japan and South Korea, as they were presented with an economically sound route for transporting their goods to Europe.

Building on this success, discussions regarding additional railroads and a new Eurasian Land Bridge have been held regularly between China, its CA counterparts and Russia. However, the most comprehensive plan, which goes beyond a single railroad project and aims to create an infrastructure network and intensify the mutual integration of regional economies, was formulated only in 2013, when the Chinese government proposed the construction of the "Silk Road Economic Belt" (consisting of six economic corridors, of which the Eurasian Land Bridge, China–Central Asia–West Asia, and the "Twenty-First Century Maritime Silk Road" are relevant to CA) OBOR/BRI concept.[50] The Asian Infrastructure Investment Bank (AIIB) that was established as an alternative to regional and global financial institutions, such as the Asian Development Bank or the World Bank (in partnership with the BRICS – Brazil, Russia, India, China and South Africa – New Development Bank, the Silk Road Fund and the SCO Interbank Association), will serve as the financing arm of the OBOR/BRI.[51] In addition to the economic benefits of this project, the Chinese government aims to address the lack of trust among some of its smaller neighbors (which was recognized in the early 1990s and described in the first section of this chapter) by offering infrastructure development, which sends a message to CA states that China is genuinely interested in contributing to their development.[52] Such efforts are designed to further strengthen the "soft-power" potential of China in Central Asia by offering a non-coercive, non-military (non-security focused) approach.

The OBOR/BRI initiative represents an increasing effort to shape positive Chinese involvement in CA that constructs a sense of common belonging. Such efforts can be considered part of China's response to the calls from regional states to implement wider integration and "to create common political, economic and informational space and to instill in the peoples of the six nations a sense of a shared destiny".[53]

On many occasions, Chinese officials have emphasized that OBOR/BRI is not an "expansion of China" but rather benefits all parties.[54] As described above, China indeed did not contemplate a major economic offensive in CA in its initial engagement with the region. However, many in CA feel that the announced goal of improving the livelihood of people in the OBOR/BRI area could be threatened by the economic and cultural expansion of an economically, politically and demographically superior power (China) in this region.[55]

Thus, new Chinese initiatives aimed at reviving the Silk Road as an area of economic cooperation, although attractive in economic terms, raise concerns among the CA states. Initiatives toward cooperation along the Silk Road involve elements such as developing transportation infrastructure (pipeline, railroad and highway construction) and enhancing trade (supported by currency swap agreements). In principle, these economic initiatives are needed. However, in terms of economic structures and the capabilities of states, many regard these initiatives as largely benefiting China, using the resources and territory of the smaller CA states but producing very marginal growth- or income-generating effects for them.[56] In particular, experts suggest that previous transportation infrastructure development designed to transport CA oil and gas to China also paved the way for the expanded penetration of cheap Chinese consumer goods into the CA region, leaving little opportunity for local production capacities to develop.[57]

In addition, some of the projects initiated under the scheme have the potential to alienate smaller countries and thus ignite intra-regional rivalry in CA, as exemplified by the railroad construction from Uzbekistan to China through Kyrgyzstan. This project connects the Uzbek city of Andijan and the Chinese city of Kashgar, with the route running though Kyrgyz Osh and Irkeshtam. This is the shortest route from China to Uzbekistan, and both countries are interested in its construction.[58] Interestingly, this route was discussed between the governments of China and Uzbekistan in 1992 during the visit of the then foreign minister of China Qian Qichen to Uzbekistan. In 1994, Uzbek president Islam Karimov spoke of the importance of constructing a direct railway from Uzbekistan to China through Kyrgyzstan during PM Li Pen's visit to Uzbekistan.[59] While China has for years been interested in several transport corridors that would connect it with other markets in Europe through CA's transport networks, this railroad is of particular interest and importance to Uzbekistan. In 1998, China signed an agreement with Uzbekistan and Kyrgyzstan to begin construction of this railroad. The agreement allows China to shorten the transportation route for goods and to avoid using Kazakh railroads. In addition to the fact that the use of Kazakh railroads results in a longer transportation time and higher costs for China, Chinese experts were also concerned about overdependence on Kazakhstan for transporting their goods. Thus, this new railroad offers China alternatives if relations with Russia or Kazakhstan worsen in the future.

In 2012, Kyrgyzstan drafted its own railroad project along this route. However, for both China and Uzbekistan, Kyrgyzstan's plans would result in a longer transportation time for their cargo and much higher costs for the project in general, and they therefore appear to be difficult for China and Uzbekistan to accept.[60] Thus, what is convenient for China might incur the displeasure of both Kyrgyzstan and Kazakhstan.

This structure of economic interactions led to a paradoxical situation: the higher the levels of interdependence with and penetration into these regions by China, the greater the concerns expressed by local governments and business communities about the possibility of Chinese domination affecting the economic development of the smaller member states.[61] This situation again calls for a proper conceptualization of the OBOR/BRI initiative that reflects both Chinese interests and the long-term interests of all CA states.

In terms of Japanese involvement, similar calls for a more mutually beneficial structure of relations were heard from CA countries during the recent visit of Japan's PM to CA from October 22–28, 2015. The visit of PM Abe to Central Asia can be termed historic because it was the first time a Japanese PM had visited all five Central Asian states. PM Koizumi visited CA in 2006, but his visit was limited to Uzbekistan and Kazakhstan. PM Abe's visit built upon previous Japanese engagement strategies as mentioned above. Japanese experts generally attempted to differentiate Japanese engagement in CA from Chinese engagement by arguing that the Chinese "assistance" provided to CA was largely a gesture of goodwill, like a "firework shoot" before the launch of major infrastructure-related projects.[62] This argument was supported by the earlier Chinese policy discourse regarding ODA being no more than a symbol of good-will.[63] Thus, in the Japanese official and academic discourse, the Japanese involvement in this area referred to as the Silk Road region in PM Hashimoto's initiative was framed as a strategy that went beyond infrastructure construction and instead attempted to transfer technology and knowledge as demonstrated in the next chapter.

South Korea has also attempted to use opportunities created by the Chinese OBOR/BRI initiative for Korean gain. However, it should be noted that South Korean Silk Road Diplomacy was launched earlier than the articulated OBOR/BRI initiative. Thus, it represents a branding umbrella scheme for expanding South Korea's corporate interests in this region. By the time South Korea's Silk Road Diplomacy was announced, South Korea's economic presence in CA, in particular in Kazakhstan and Uzbekistan, was significant in terms of ODA assistance, direct investments and human resource development. In terms of aid disbursement, Uzbekistan ranks the highest in the region, accounting for a third of the assistance volume, while Kazakhstan is second highest, accounting for approximately 6 percent. South Korean assistance and investment programs, in addition to considering the importation of resources from CA to South Korea (similar to the 2008 contract to import 2,600 tons of uranium from Uzbekistan), also aim to link CA resources to international markets. In this way, these programs aim to establish economically sustainable production, extraction and reproduction cycles that are marketable in CA and beyond the region and that benefit South Korean conglomerates.

Park Geun-hye's Eurasia Initiative, in October 2013, aimed to further conceptually develop South Korean engagement in this region and called for linking energy and logistics, as explained in the previous section. In attempting to construct an engagement of its own, South Korea is not counterposing this initiative to the Chinese OBOR/BRI initiative, thus again supporting the point that there is no structural rivalry and competition between the Chinese, South Korean and the Japanese initiatives either at the level of discourse or practice. Some even linked such South Korean behavior – which shows signs of being a hybrid of pragmatic opportunity-seeking with idealizing and ambitious goal setting – to the notion of "middle-power diplomacy" interpreted as "a search for conceptual breakthroughs by political leaders and policy intellectuals, with the aim of elevating the country's place in the world and enabling more proactive diplomatic roles".[64] As they conceive it, the weakness of South Korean approaches (which to some extent can also be applied to explaining the Japanese policies in the CA region) is that the concepts forwarded by such "middle-power" states fall short of articulating a clear, longer-term strategic vision linked to coherent policy practices.[65]

Conclusion

As is demonstrated in various parts of this chapter, the notion of the Silk Road has been successfully utilized by a number of states in their engagement with the CA region. Although in foreign policies the Silk Road is presented as a static concept, this chapter argued that the Silk Road as a term has come to represent the diverse concepts of engagement of a number of powerful states vis-a-vis the CA region. This article also demonstrated that the "Silk Road"-branded strategies of these states have initially been reactions to the collapse of the Soviet Union and attempts to minimize the negative effects. They have later grown into initiatives seeking to incorporate CA countries within the foreign policies of China, Japan and South Korea. Such attempts to place the Silk Road into one's foreign policy are undergoing constant evolution and social construction depending on international environment and constraints. As demonstrated by the policies of China, Japan and South Korea, the Silk Road is not a foreign policy doctrine for these states but rather a discursive strategy of engagement that largely exists in the realm of narration. This narration is also a matter of social construction that is subject to change depending on the international environment of the country (China, Japan, Korea, etc.) that produces such narratives, the alternative narratives that compete for wider international acceptance and the country's vision of "self" and the "other" in the international context.

In a similar manner, Silk Road strategies are an attempt to both define "self" (what Japan, China and Korea stand for) and socially construct the image of "others" to win the hearts and minds of targeted states by demonstrating how the (Chinese, South Korean and Japanese) "self" varies and is much more advantageous than alternative "others". In this sense, as demonstrated in this chapter, China, Japan and South Korea had no predetermined final goals or aims in their Silk Road strategies from the start. Rather, these states re-constructed and re-shaped their strategies according to changing perceptions and discursive categories of "self", "identities", "values", "threats" and "opportunities".

As is demonstrated by the discussion of each "Silk Road", this rhetoric has been selected because it is one of the few available that can emphasize the common history, heritage, values and identity of such diverse countries. The "Silk Road" is also frequently referred to as an engagement strategy because it projects the image of decolonization and reflects the agency of CA states as independent actors, which is easy for the target states to comprehend and thus accept.

Finally, frequently, the "Silk Roads" are depicted as having no similarities with each other. Although their distinctions are obvious, this chapter has also demonstrated that there are certain similarities in the logic of their construction and in the discursive expressions that are meant to be the "selling points" of these initiatives. In the cases given, these related to how China, Japan and Korea depict their genuine interest in contributing to the development of the CA region, how they create historical linkages to modern-day Silk Road projects and how they emphasize the "win–win" structure of their relations. However, these represent elements of coercion that are meant to ensure a better reception of the goals and objectives of the diverse "Silk Roads" envisioned by these powerful states.

In addition, the areas of energy research development and transportation networks always feature as of high potential. However, these areas are attributed different meanings and importance. While for South Korea, Silk Roads may imply construction of railroad infrastructure, for China, this type of infrastructure is no more than a tool in achieving ambitious tasks in the areas of security, political stabilization and economic expansion. The dynamism in cooperation along these areas differs depending on the geographic proximity of each country. For China which shares borders with CA states, the logistics of such cooperation are less problematic than they are for South Korea and Japan.

Notes

1 Timur Dadabaev, "Japan's Search for Its Central Asian Policy: between Idealism and Pragmatism", *Asian Survey* 53 (2013): 506–532, doi:10.1525/as.2013.53.3.506.
2 A Cooley, "New Silk Route or Classic Developmental Cul-de-Sac? The Prospects and Challenges of China's OBOR/BRI Initiative", *Ponars*, July 15, 2015, www.ponar seurasia.org/node/7833.
3 M Li, *Zhōng yà Yan'zyu* [Research into CA] (Lanzhou University, 1995); G Xing, *Zhungo sin duli de zhōng yà godzya guansi* [Relations of China with newly independent states] (Harbin, 1996); W Pei, *Zhōng yà wǔ guó gàikuàng* [The general situation of five CA states] (Urumqi: Xinjiang People's Publishing House, 1997); G Xing, C Zhao and Z Sun, *Zhōngguó yui Zhōng yà* [China and Central Asia] (Beijing, 1999).
4 Z Wang and B Ding, *Zhōng yà guójì guānxì shǐ* [History of international relations with Central Asia] (Beijing: Hunan Press, 1997).
5 S Cornelissen and I Taylor, "The Political Economy of Chinese and Japanese Linkages with Africa: A Comparative Perspective", *The Pacific Review* 13, no. 4 (2000): 615–633.
6 S Zhao, "Foreign Policy Implications of Chinese Nationalism Revisited: The Strident Turn", *Journal of Contemporary China* 22, no. 82 (2013): 535–553.
7 G Xing, *Zhongguo he xin du li di Zhong Ya guo jia guan xi* [Relations of China with newly Independent States of Central Asia] (Ha'erbin: Heilongjiang jiao yu chu ban she, 1996).
8 Xing, *Zhongguo he xin du li di Zhong Ya guo jia guan xi*; G Xing, C Zhao and Z Sun, *Zhōngguó yui Zhōng yà*.
9 Xing, Zhao and Sun, *Zhōngguó yui Zhōng yà*, pp. 155–170; Xing, *Zhongguo he xin du li di Zhong Ya guo jia guan xi*, p. 101.
10 Xing, Zhao and Sun, *Zhōngguó yui Zhōng yà*, pp. 135–140.
11 Xing, Zhao and Sun, *Zhōngguó yui Zhōng yà*, p. 212.

12 Xing, Zhao and Sun, *Zhōngguó yui Zhōng yà*, p. 209.

13 Rosita Dellios, "Silk Roads of the Twenty-First Century: The Cultural Dimension", *Asia & the Pacific Policy Studies* 4, no. 2 (2017): 225–236, doi:10.1002/app. 5.172.

14 Shanghai Cooperation Organization (SCO) [Shànghǎi hézuò zǔzhī], *Xīn ānquán guan yui xīn jīzhì* [New thinking and new mechanism for security] (Beijing: Current Affairs Press, 2002).

15 Z Fan, *Zhōng yà dìyuán zhèngzhì yǔ wénhuà* [Central Asia's geopolitical culture] (Luluqi: Xinjiang People's Publishing House, 2003).

16 Z Sun, *Zhōng yà qíngshì hé dìqū ānquán* [State of affairs in Central Asia and regional security] (Beijing, 2001); Z Sun, *ZhongYa Xingeju yu Diqu Anquan* [The New Central Asian Order and Regional Security] (Beijing: China Social Sciences Press, 2001); Z Sun, "The Relationship between China and Central Asia", *Slavic Eurasian Studies* 16, no. 1: 41–63, http://src-h.slav.hokudai.ac.jp/coe21/publish/no16_1_ses/03_zhuangzhi.pdf.

17 J Xùe and G Xing, *Zhōngguó yǔ zhōng yà* [China and Central Asia] (Beijing: Social Sciences Literature Publishing House, 1999).

18 R Hashimoto, Address to the Japan Association of Corporate Executives, Tokyo, July 24, 1997, www. Japan.kantei.go.jp/0731douyuukai.html.

19 Hashimoto, Address to the Japan Association of Corporate Executives.

20 Dadabaev, "Japan's Search for Its Central Asian Policy"; Timur Dadabaev, "Chinese and Japanese Foreign Policies towards Central Asia from a Comparative Perspective", *The Pacific Review* 27, no. 1 (2014): 123–145. doi:10.1080/09512748.2013.870223.

21 Hashimoto, Address to the Japan Association of Corporate Executives, p. 2.

22 T Uyama, C Len and T Hirose, eds., *Japan's Silk Road Diplomacy: Paving the Road Ahead* (Washington, DC: Central Asia-Caucasus Institute and Silk Road Studies Program, 2008), p. 111.

23 Hashimoto, Address to the Japan Association of Corporate Executives, pp. 2–3.

24 A Kawato, "What Is Japan Up To in Central Asia?", In *Japan's Silk Road Diplomacy: Paving the Road Ahead*, ed. C Len, T Uyama and T Hirose (Washington, DC: Central Asia-Caucasus Institute and Silk Road Studies Program, 2008).

25 "Chuou Ajia hatten no kokusaiteki jyoken to Nihon" [Development of Central Asia: International conditions and Japan], *Gaiko* 34 (2015): 21–34, especially pp. 31–33.

26 Hashimoto, Address to the Japan Association of Corporate Executives, p. 3.

27 Hashimoto, Address to the Japan Association of Corporate Executives, p. 10.

28 U Makhmudov, "Sengo Nihon no Chuou Ajia Seisaku to Senryaku: Chuou Ajia plus Nihon Taiwa wo Chushinni" [CA policy and strategy in post-War Japan: The case of CA plus Japan Dialogue], *Housei Daigaku Daigakuin Kiyou* 77 (2016): 65–90 (see especially p. 83), http://repo.lib.hosei.ac.jp/bitstream/10114/12735/1/grad_77_mahmudov.pdf.

29 J Sato, "A Japanese Approach to Assistance: Cherishing the Recipient Experience", *Asahi Shimbun*, October 31, 2011, www.asahi.com/shimbun/aan/english/hatsu/eng_hatsu111031.html, accessed on September 15, 2012; F Furuoka, "A History of Japan's Foreign Aid Policy: From Physical Capital to Human Capital", 2007, Munich Personal RePEc Archive, Chapter No. 5654.

30 B Hwang, "A New Horizon in South Korea-Central Asia Relations: The ROK Joins the 'Great Game'", *Korea Compass*, December 2012, pp. 1–7.

31 Joungho Park, *The Role of Korea–Central Asia Cooperation Forum for Eurasia Initiative* (Sejong: Korea Institute for International Economic Policy, March 22, 2016), www.kdevelopedia.org/Resources/economy/role-korea-central-asicooperation-forum-eurasiinitiative-04201603240143879.do?fldIds=TP_ECO%7CTP_ECO_EA#.WhJYz kpl-Uk, accessed 23 November, 2017.

32 H Shin, "Office Launched for Korea–Central Asia Cooperation", *The Korea Herald*, July 11, 2017, www.koreaherald.com/view.php?ud=20170710000880, accessed November 23, 2017.

33 Hwang, "A New Horizon", p. 3.
34 MB Lee, *New Asia Initiative* (2009), www.mofat.go.kr/mofat/pcrm/eng5.doc, accessed November 22, 2017.
35 Hashimoto, Address to the Japan Association of Corporate Executives, p. 4.
36 Hashimoto, Address to the Japan Association of Corporate Executives, p. 4.
37 Interview with FM Fumio Kishida, "Shinrai to sougo kanshin no kiban no ueni" [Building on the foundations of trust and mutual interest], *Neutral Turkmenistan*, May 9, 2017, p. 3, in Japanese and Russian at www.mofa.go.jp/mofaj/p_pd/ip/page4_002989.html (translated by the author).
38 Xue and Xing, *Zhōngguó yǔ zhōng yà*, pp. 102–106.
39 Xue and Xing, *Zhōngguó yǔ zhōng yà*, pp. 127–130.
40 SCO, *Xīn ānquán guan yui xīn jīzhì*, pp. 9–11.
41 SCO, *Xīn ānquán guan yui xīn jīzhì*, pp. 9–11.
42 Xing, *Zhongguo he xin du li di Zhong Ya guo jia guan xi*, pp. 152–158.
43 Sojung Yoon, "President Emphasizes Cooperation with Central Asia", *Korea.net*, July 8, 2014, www.korea.net/NewsFocus/policies/view?articleId=120478, accessed on November 19, 2017.
44 Matteo Fumagalli, "Growing Inter-Asian Connections: Links, Rivalries, and Challenges in South Korean–Central Asian Relations", *Journal of Eurasian Studies* 7, no. 1 (2016): 39–48 (especially p. 45), doi:10.1016/j.euras.2015.10.004.
45 Hwang, "A New Horizon", pp. 1–7.
46 Yoon, "President Emphasizes Cooperation with Central Asia".
47 Fumagalli, "Growing Inter-Asian Connections"; T Kim, "Beyond Geopolitics: South Korea's Eurasia Initiative as a New Nordpolitik", *Korea National Diplomatic Academy*, February 16, 2015, www.theasanforum.org/beyond-geopolitics-south-koreas-eurasia-initiative-as-a-new-nordpolitik/, accessed November 22, 2017.
48 KE Calder and V Kim, "Korea, the United States, and Central Asia: Far-Flung Partners in a Globalizing World", KEI (Korean Energy Institute) Academic Paper Series 3, no. 9 (December 2008), https://ja.scribd.com/document/69499081/Korea-the-United-States-and-Central-Asia-Far-Flung-Partners-in-a-Globalizing-World-by-Kent-E-Calder-and-Viktoriya-Kim.
49 Xing, *Zhongguo he xin du li di Zhong Ya guo jia guan xi*, p. 168.
50 Office of the Leading Group for the Belt and Road Initiative, *Building the Belt and Road: Concept, Practice and China's Contribution* (Beijing: Foreign Languages Press, May 2017), pp. 11–17, https://eng.yidaiyilu.gov.cn/wcm.files/upload/CMSydylyw/201705/201705110537027.pdf, accessed March 12, 2018.
51 National Development and Reform Commission, Ministry of Foreign Affairs and Ministry of Commerce of People's Republic of China, "Vision and Actions on Jointly Building Silk Road Economic Belt and 21st-Century Maritime Silk Road", March 30, 2015.
52 Ministry of Foreign Affairs of the People's Republic of China, "Joint Declaration on New Stage of Comprehensive Strategic Partnership Between the People's Republic of China and Republic of Kazakhstan", August 31, 2015, www.fmprc.gov.cn/mfa_eng/wjdt_665385/2649_665393/t1293114.shtml, last seen on January 25, 2018).
53 B Nurgaliev, Statement of the SCO Secretary-General Bolat K Nurgaliev at the Security Forum of the Euro-Atlantic Partnership Council, Astana, June 25, 2009, www.sectsco.org/EN/show.asp?id=104.
54 "Posol: proekt 'Odin poyas, odin put' – eto ne ekspansiya Kitaya" [Ambassador: "One Belt, One Road" – is not expansion of China], *Podrobno.uz*, May 29, 2017, www.podrobno.uz, last accessed on May 29, 2017.
55 M Auezov, "Ex-Ambassador of Kazakhstan to China Concerned over China's Classified Documents", *Tengri News*, 2015, www.en.tengrinews.kz/politics_sub/Ex-Ambassador-of-Kazakhstan-to-China-concerned-over-Chinas-21645/, last accessed on January 15, 2016; M Auezov, "Kazakhstan Must Stop Wavering between Russia and China to Pursue Central Asian Consolidation", *Interfax*, January 29, 2013.

56 "Stroitel'stvo zh/d Chuy-Ferghana vygodnaya al'ternativa doroge Kitai-Kyrgyzstan-Uzbekistan: expert" [Construction of the railroad Chuy-Ferghana is beneficial alternative to the road of China-Kyrgyzstan-Uzbekistan: expert], *KyrTag (Kyrgyz Telegraph Agency)*, April 20, 2012, www.kyrtag.kg/?q=ru/news/19472, accessed on 20 April 2012.

57 Ablat Khodzhaev, *Kitajskij Faktor v Tsentral'noi Azii* [Chinese factor in Central Asia] (Tashkent: Fan, 2007), pp. 69–72.

58 A Titova, "Uzbekistan hochet postroit Andijan–Osh–Irkishtam–Kashgar" [Uzbekistan wants to build Andijan–Osh–Irkishtam–Kashgar], *Kloop*, September 9, 2017, https://kloop.kg/blog/2017/09/09/uzbekistan-hochet-postroit-trassu-andizhan-osh-irkeshtam-kashgar/, accessed 23 September, 2017.

59 Khodzhaev, *Kitajskij Faktor v Tsentral'noi Azii*, p. 103.

60 Bruce Pannier, "No Stops in Kyrgyzstan for China-Uzbekistan Railway Line", *Radio Free Europe/Radio Liberty*, September 3, 2017, www.rferl.org/a/qishloq-ovozi-kyrgyzstan-uzbekistan-china-railway/28713485.html, accessed 23 September, 2017.

61 Auezov, "Ex-Ambassador of Kazakhstan to China".

62 "Chuou Ajia hatten no kokusaiteki jyoken to Nihon".

63 Xing, *Zhongguo he xin du li di Zhong Ya guo jia guan xi*, pp. 104–106.

64 S Kim, *South Korea's Middle-Power Diplomacy: Changes and Challenges*, Chatham House Research Papers (London: Chatham House, 2016), www.chathamhouse.org/sites/files/chathamhouse/publications/research/2016-06-22-south-korea-middle-power-kim.pdf, accessed November 22, 2017.

65 Kim, *South Korea's Middle-Power Diplomacy*.

3 Discourses of rivalry or rivalry of discourses?

In recent years, the Central Asian (CA) region has been a major focus of attention for many regional and global players. In late 2013, President Xi Jinping of China visited CA and Southeast Asia and announced China's commitment to launching the Silk Route Economic Belt to connect China and the countries of CA and Western Asia and to boost cooperation and the Chinese presence in the region. This initiative is part of the Chinese One Belt, One Road (OBOR) initiative, also interchangeably referred to as the Belt and Road Initiative (BRI). For China, this initiative has many aspects. On the one hand, it is a framework to develop connectivity and cooperation among countries in Central Eurasia.[1] On the other hand, it is also related to the global strategy to connect Chinese goods and producers to the markets in a more efficient way, which has relevance to sub-national policies.[2] In addition, this initiative is part of a domestic policy intended to bring development to various less-developed parts of China, such as Xinjiang.[3] As one aspect of OBOR/BRI, in 2014, China announced the establishment of the China-led Asian Infrastructure Investment Bank (AIIB).[4] The AIIB has a focus beyond the OBOR/BRI initiative, although many OBOR/BRI member states became AIIB members in the hope of receiving financial support for their participation. This initiative, paired with the Shanghai Cooperation Organization (SCO) initiative, represents an attempt by China to assert itself on a global scale as an active agenda-setting power and to expand its role in various areas.[5]

Japanese engagement has also shown a dynamism rarely seen in previous years. On October 22–28, 2015, Abe Shinzo, the prime minister (PM) of Japan, visited CA. This was only the second visit by a PM of Japan to the region (the first was the visit of PM Koizumi Junichiro in 2006) and represented the first visit of the Japanese PM to all five CA states. It built upon previous Japanese official development assistance (ODA) projects, Japan's engagement strategies and the "Central Asia plus Japan" (CAJ) initiative announced in 2004. During the visit, the Japanese PM, accompanied by representatives of Japanese companies and educational institutions, attempted to cement the presence of the Japanese business community in CA, as exemplified by contracts signed during the visit for the joint exploration of gas fields in Turkmenistan, Uzbekistan and Kazakhstan. In addition, Japan aims to boost its educational and humanitarian involvement in this region to support both its "Cool Japan" policy and its broader soft-power construction in this region. Although Japan does not directly link these initiatives with the intensification of Chinese initiatives in this region, many observers connect them to (real or imaginary) Japanese attempts to secure access to CA for Japanese corporate and state institutions and, by doing so, to offer sources of funding and development initiatives as alternatives to those offered through Chinese schemes.[6]

The Chinese and Japanese visits and proposed initiatives mentioned above occurred during roughly the same period, leading many analysts and experts to believe that these policies were related to each other in a cause-and-effect relationship. Therefore, as demonstrated below, many argued that the Japanese and Chinese expansions of these countries' CA engagements are the result and consequence of competition for regional mineral resources and influence. In addition, some experts claim that such competition implies a mutually exclusive rivalry between Japan and China, with each aiming to limit the capacity of the other in this region. They warn that China's "expansion into the region entails strategic consequences that may severely challenge Japanese foreign policy and security".[7]

Although the intensity of the visits and initiatives proposed for CA emphasizes the importance attributed to this region by China and Japan as a "new frontier" for their foreign policy engagement, it remains unclear whether it signifies the outcomes of rivalries between these powers or simply an attempt to secure their interests regardless of the actions of the other nation. In addition, the question of how these initiatives are perceived by the CA public and leadership remains open. The impact of friction between China and Japan in light of their East Asian experiences does not shed much light to help understand their involvement in CA. To fill the academic gap, this chapter raises the following questions: How can the intensification of Japanese and Chinese foreign policy engagements in this region be interpreted? Do these engagements relate to rivalry between the two countries or alternative framings and justifications of their foreign policies in this part of the world? How are these initiatives perceived by the CA public and leadership? What are the images and expectations associated with Japan and China in light of the recent intensification of the foreign policies of these states?

This chapter argues that the discourse of competition for natural resources and domination in CA between China and Japan is premature and empirically unproven.

To put it into comparative perspective, there are only 18 Japanese companies operating in Uzbekistan, which is the demographically largest country in CA, compared to 410 Korean and 480 Chinese companies in 2016.[8] China is also a top trading partner for resource-rich Uzbekistan, with a trade share constituting 18 percent of its international trade, surpassing even Russia. Japan is not within the top ten countries. Additionally, Chinese contracts for energy resources feature prominently in the road maps of cooperation (reaching billions of dollars), exemplified by its dealings with Uzbekistan in 2017.[9] In contrast, Japan largely exports to CA states machinery and industrial goods (13.56 billion yen to Uzbekistan, 30 billion yen to Kazakhstan, 662 million yen to Tajikistan and 2.4 billion yen to Kyrgyzstan), while importing textile yarn, fabrics and nonferrous metal (from Uzbekistan for 500 million yen), radioactive material and nonferrous metals (from Kazakhstan for 141.1 billion yen), and fruits and non-metallic ware (from Tajikistan for 163 million yen and Kyrgyzstan for 153 million yen).

Thus, it is important to provide a note on the importance of the energy resources of CA for Japan as an area of potential conflict and rivalry with China. Discursively, the exploration of energy resources (to include rare metals, oil and gas) has always been an important pillar that Japan framed as an attempt to diversify its energy dependency on the Middle East. However, the importance of energy supplies from CA has been overstated due to logistical problems related to delivering these resources, which include but are not limited to the geographically distant nature of the region and a lack of access to sea ports in the regional states, all of which prevent the construction of a proper infrastructure for delivering CA energy resources to Japan.[10] Additionally, in terms of geopolitical location, countries that are sandwiched between CA states and Japan (namely, China, Russia and South Korea) cannot be considered Japan-friendly in terms of their foreign policies, making it more difficult to construct energy-related infrastructure from CA to Japan. Thus, although this theme has featured prominently in the FM Aso 2006 "Arc of Freedom and Prosperity" speech and the 2015 PM Abe visit to CA (particularly in Kazakhstan and Turkmenistan), the rhetoric of the importance of energy resources to Japan largely remains in the realm of rhetoric and produces little in terms of tangible outcomes. The few examples of practical results of cooperation include the Kazakhstan–Japan agreement on the joint exploration and use of mineral resources that coincided with a radical decrease in the supply of rare metals from China to Japan following a 2010 boat incident between the two countries.[11] This agreement supported previous corporate plans, such as those signed by Japan's Kansai Electric Power Company (KEPCO) with KazAtomProm and contracts between Itochu and KazAtomProm to develop uranium deposits. According to the terms of this agreement, Kazakhstan could provide up to 25 percent of the Japanese demand for uranium within the next decade. Similarly, based on the visit of PM Abe to CA, a few contracts have been signed for effecting the participation of Japanese companies in the construction of processing plants in CA (particularly in Kazakhstan and Turkmenistan). However, these initiatives are not connected to transporting energy resources to Japan but rather represent the corporate participation of Japanese companies in processing the energy resources for further exports to China and other states.[12] Thus, the discourse of the importance of using CA regional energy resources as alternative sources of energy represents more wishful thinking than a realizable and practical goal. The statistics on exports from CA to Japan and cooperation road maps[13] also demonstrate that energy resources do not feature prominently in the trade between this region and Japan, again supporting the points above.

The more active involvement of these countries in CA in recent years is rooted in the changing international identities of these states and the fact that both have demonstrated a relatively moderate involvement in CA in previous decades compared to other states, such as Russia.

This chapter further argues that the overlapping areas in which both Japan and China operate primarily reflect a process of reconstructing their foreign policy engagements globally/internationally in accordance with changing internal and international environments, as opposed to attempts to compete with one another.

This chapter also emphasizes the differences in approaches to CA. Discourses used by China and Japan are confined within a similar framing of both having "non-Western" and "Asian" features enabling them to better understand the concerns of CA states. At the same time, their strategies differ somewhat in content and approach. The Chinese idealist rhetoric of the common benefits of cooperation is utilized to achieve the pragmatic goals set in the OBOR/BRI initiative and the like, whereas Japan's foreign policy displays signs of transition. For a long time, Japan used its dual identity of having "Asian" roots and committing to "universal" standards and values in its foreign policy behavior to build its relations with CA states. This duality used as an advantage of the Japanese approach is analyzed in Chapter 7 of this study.

However, as seen from the agenda of PM Abe's visit to CA, Japanese foreign policy is showing signs of departing from such dualism with certain shifts toward more pragmatic and functional goal setting (energy resources, joint-venture development and soft-power construction). Such pragmatism is becoming a social norm and value upon which common vision and perception is built in various engagement schemes (whether led by China, Japan or Russia).

Rivalry and great-power politics in CA international relations

As mentioned in the introductory chapter, the notion of a possible rivalry between China and Japan in CA relates to the general tendency among experts to connect any activization of the foreign policy of powerful states in CA to the notion of inevitable rivalry, competition and a new Great Game.[14] However, such a framing limits discussion of the possibility, however remote, that the interests of these states may overlap, merge, mutually construct each other and even, at times, be shared. This tendency also derives from the vision of international relations through the lens of realist (whether classical, structural or offensive realist) assumptions. Accordingly, inter-state relations in CA have been described as developing along the lines of anarchy, selfishness, egoism, the absence of international order, and the importance of increasing the capability to protect and expand national interests in this part of the world. National interests are also defined in material terms of resources, territory and financial gains.

Interpretations of the motivations and behavior of various powers in CA have centered around the theme of domination and rivalry ever since CA countries gained their independence in 1991.[15] Initial discussions focused on the geographical positioning of CA in close proximity to Russia, thus terming it "Russia's backyard", and the fact that the Russian government lacked the resources to enforce its ambitions in CA in the early 1990s.[16] The diversification of CA states' relationships with states other than Russia resulted in discussions of challenges to Russian domination and attempts to fill the gap left by Russia in "withdrawing" from this region. As part of this discussion, any attempts by the EU and the US to engage these states resulted in an imaginary resurrection of the Great Game in this region that involved not only large powers but also smaller regional states "playing" their "games"[17] or attempting to develop their own programs in line with those of great powers.[18] Similarly, the rise of states such as China and India resulted in references to great-power Great Games.[19] Even liberal ideas related to the spread of democracy and human rights were often regarded as a tool of Western penetration aiming to constrain and pressure the Russian sphere of influence.

In a post-9/11 world, the rhetoric of US–Russian rivalry in this region has further intensified.[20] This rhetoric either portrays an open confrontation between Russia and US-led Western states or emphasizes the important detrimental impact of larger powers such as Russia and the US on establishing peace and stability in Afghanistan. In both interpretations, minor roles, if any, are attributed to CA states, with their only option being to "get on the bandwagon" of the dominating power. For example, according to the depiction of CA support of the US-led Afghan campaign from this point of view, the US is expanding its influence in this region, while Russia has little power to oppose that expansion. In such an interpretation, CA states are seen as joining the US. This narrative further depicts the color revolutions of 2005 in former Soviet territories (Georgia, Ukraine, Kyrgyzstan, etc.) as threatening the existence of CA regimes, forcing them to join with Russia to survive and thereby limiting the ability of the US to expand its power in the region. US military bases have been closed, while Russia has increased its involvement by attempting to re-build its area of influence through constructing the Customs Union and the Eurasian Economic Union and setting the goal of creating the Eurasian Union.[21]

Beginning in 2001, this narrative of US–Russian rivalry in CA was further illuminated by the rise of China and the impact of the SCO on this region.[22] The narrative has been characterized by China's attempts to expand its economic power in this region and secure its political standing against any attempts by the US and even Russia to encroach on issues of vital concern for China.[23] In the latter half of the 2000s, with the decline of US and EU power in this region, the discussion has been dominated by how Russia and China will oppose US interests in this part of the world and simultaneously reconcile their differences over how the region should be developed.[24] With such a "Western decline", some authors have called for reviving the EU and US influences in this region, claiming that "the Central Asian states will continue to see Western recognition as a welcome balance to Moscow and Beijing".[25] In these interpretations, the role of getting on the bandwagon is again attributed to CA states, while China and Russia balance against the US or the EU and at times all these states balance against each other.[26] With the announcement of the Chinese OBOR/BRI initiative, an increasing rhetoric of Chinese peaceful expansion makes an additional contribution to the typical narrative of power politics in this region.[27] Some have even questioned the economic rationality of OBOR/BRI, emphasizing its political benefit solely for China.[28]

While Japanese ambitions and influence in this region have never been as strong as those of Russia and China, every visit by a high-ranking Japanese official, such as a foreign minister (FM) or PM, and any announcement of an intensification of Japanese foreign policy in this region have inevitably been linked to the perceived global rivalry between these states for domination.[29] Such references are often made in Japanese and foreign media outlets.[30] For instance, media coverage featured powerful images of the "new Great Game", "Japanese revenge", the "expansion of Japanese influence" or "Japanese play in resource diplomacy and the great game".[31] Frequently, Japanese initiatives are viewed as attempts to regain "the lost initiative to Russia and China in relations with CA countries"[32] or as in line with "Japanese–Chinese rivalry in the third world".[33] Even experts who accept that the Great Game is no longer possible in CA warn that co-exploitation of this region by Russia and China is a matter of concern and that Japan offers an alternative to these great powers.[34] References and depictions similar to those described above were applied to Japanese foreign policy with every new initiative undertaken by Japan in this region, such as PM Hashimoto's Silk Road Diplomacy (1997), the CAJ Forum (2004), Japanese PM Koizumi's visit (2005), FM Aso's Arc of Freedom and Prosperity (2006) and PM Abe's visit (2015).

However, this chapter argues that realist arguments (the interests of states defined as a constant rivalry for survival and domination, national interests defined in material terms, and anarchy and gains defined in relative and not absolute terms) about the CA region are neither empirically proven nor logically consistent. The weaknesses of such arguments about the inevitable and constant rivalry in CA are three-fold. First, such depictions show the foreign policy behavior of states as static and unchanging. The socially constructed nature of their relations and mutual relevance and the process of othering in defining "self" are not attributed any significance.[35] However, the policies of the US, Russia, China and Japan have undergone various movements with sometimes drastic changes in how they engage with CA. They also have a certain relevance to each other in defining competitive advantage. These discursive dynamics remain un-accounted for by a static imagination of rivalry. Second, such depictions attribute equal roles to all powerful states – the US, EU, Russia, China or Japan – and suggest that all these states seek to dominate the CA region. However, judging from official statements and contracts signed during the visits of senior officials to this region, the interests and goals of these powerful states are much more diverse than simple domination and control. The third weakness of the traditional arguments of domination is that they attribute no actor-agency to CA states, instead depicting them as absorbers of the policies of various powers. The domination narratives either display CA as being dominated by larger powers or rush to the alternative extreme of depicting CA states as adapting to the US, Russia and China or "playing" them against each other to secure greater gains. In contrast to such an approach, this chapter emphasizes that the interactions of states in CA cannot be explained by the logic of domination and manipulation but are rather socially constructed, depending on the types of interactions and constraints that shape their identities, and thus define the values and patterns of interactions. In this sense, the manner in which their interactions are shaped can hardly be depicted in a static way, as these interactions change with changes in the international or internal environments, threats and challenges, which then reshape the values and identities of these states.

Using the case of Chinese and Japanese engagement in CA, this chapter argues against the simplicity of the rationalist vision and instead employs constructivist rhetoric, which emphasizes changing identities and orientations and the constructed nature of foreign policy. This chapter suggests along a constructivist line that, in contrast to the positivists' explanations, it is not only the material factors and interests of states that matter in China's and Japan's motivation for engaging CA. Importantly, the motivations of these powers and other CA states are socially constructed; they are socially contingent in that they depend on interaction. In such an analysis, cultural and institutional factors establish the environment for actors in global politics. Therefore, cultural environment and identity affect states' behavioral incentives. Although, as explained above, rationalists treat the interests of China and Japan as stable, predetermined and mutually exclusive, this chapter treats those interests as changing according to changes in interaction and environment. Therefore, the great powers and CA states become who they are within these interactions, and their interests are defined by these interactions. Although the general agenda of engaging CA states coincides in the case of China and Japan, their strategies, areas of engagement and long-term plans differ and are not necessarily mutually exclusive and competitive.[36] Within such a structure, it can also be the case that the discourse of rivalry serves the purpose of the CA states, allowing them to create the image of competition and thereby further motivating the external powers to make economic and political commitments to this region, thus helping CA states to achieve their goals.

To demonstrate the validity of this framework of analysis, this chapter employs constructivist tools of analysis by first analyzing how China and Japan discursively integrate this region into internal and external policies (imagination) (the second section) and norms and concepts according to which they justify their actions in CA (communication) (the third section). It further analyzes the expectations of both Japan and China in CA; these expectations serve as both motivation and a source of constraint for engagement with CA.

At the same time, the constructivist explanation does not imply that the engagement of these countries is mutually irrelevant. As is argued throughout this chapter, both Japan and China frame their foreign policy through the lens of "otherness" by defining their engagement in this region and referring to an "other" that is different from the "self". Such efforts to define "self" through "other" are analyzed in two sections of this chapter: first in how these countries frame CA in their foreign policies and second in how this framing resonates with the general public in the section on the public image of China and Japan. In such "othering", the Japanese government and expert community focus on how China acts in the CA region so that they can define and place Japan as a "better", "more reliable" and "more genuine" partner for the region.

Framing Chinese and Japanese engagement in post-Soviet CA

As has been stated in Chapter 2 of this volume and related studies, CA represented an unknown region for both China and Japan following the dissolution of the USSR.[37] In addition to lacking a basic understanding of the foreign policies of the newly independent states, both China and Japan faced changing political, economic and international environments in the aftermath of the cold war. Therefore, their engagement strategies toward this region were largely socially constructed, reflecting the uncertainties and new international realities mentioned above. The process of the social construction of these engagement strategies can be traced through the dynamics of the constantly changing agenda setting within the initiatives of China (Shanghai Five through to SCO and on through to OBOR/BRI) and Japan (Eurasian/Silk Road Diplomacy through to CAJ). At the initial stage of their interactions, the CA region was regarded as an area that presented more challenges than opportunities.[38] Additionally, their foreign policy orientation was considered to be coordinated with Russia. Thus, the Chinese and Japanese policies for this region were drafted largely to correspond to this vision of the region.

It would be wrong to assume that the Chinese and Japanese approaches to this region were designed specifically to reflect the peculiarities of each country. Rather, they were largely in line with their general foreign policy principles, which for China focused on the goal of establishing good-neighbor relations and non-interference in each other's internal affairs. China's primary objective was to prevent any kind of possible aid from across the border to groups that the Chinese government considered "separatist". China also designed the scheme of cooperation to allow it to seek and persecute such groups and individuals on CA soil. There is a certain similarity between the Chinese and the Japanese approaches to this region in that Japan also initially attempted to approach CA through the prism of its Russian foreign policy, which in later years shifted toward integrating CA engagements into a wider framework of its Asian policy.

In the process of defining the issues and areas of cooperation, both China and Japan point to the long history of interactions dating back into the pre-Soviet period. As FM Kishida emphasized during a visit to Turkmenistan in May 2017,

> For many Japanese people, Central Asia is associated with the Silk Road, as it brought Buddhism, which is the basis of the Japanese culture, and

enriched Japanese culture by introducing civilizational and cultural influences of the West; all were communicated through the Silk Road.[39]

China also discursively uses historical connections with CA to construct mutually beneficial "win–win" relations that have received a warm welcome by CA states that feared that their relatively small economies and societies would eventually be dominated by China.[40] Building a China-friendly belt became especially important for China after the Tiananmen events of 1989 in an environment characterized by harsh criticism of Chinese governance in the US and Europe.[41] This was also the time of the shift of China's internally oriented policies toward a greater emphasis on external engagements, including in the post-Soviet space. It was both "reactive", as China was trying to adapt to the negative impact of the collapse of the Soviet Union, and "proactive", as it searched for its new place in the international community.[42]

Challenges facing Japan in the process of defining the principles and goals of its partnership with regional countries turned out to be greater than those facing China. The first Japanese effort to establish a Japanese presence in this region was the Japanese Silk Road Diplomacy launched in 1997, which became one of the first international diplomatic initiatives appealing to the connectivity and revival of the Silk Road. This was undertaken under PM Hashimoto Ryutaro's administration. Hashimoto's understanding of this region has been informed by the Obuchi Mission[43] and Hashimoto's interactions primarily with Russia. Despite launching the Silk Road/Eurasian Diplomacy, Hashimoto never traveled to the CA region and the Caucasus. This partly reflects the focus of Japanese foreign policy toward the US, as defined by its strategic alliance with it, toward China, due to Japan's economic commitments to it, and toward Russia, oriented at resolving territorial disputes. This foreign policy agenda constrained Japanese PMs' visits and did not leave much space for other regions, including CA. The Obuchi Mission traveled to Russia, Turkmenistan, Kyrgyzstan, Kazakhstan and Uzbekistan from June 28 to July 9, 1997.[44] While the mission focused extensively on Russia, it also provided ideas about how to approach CA.

Hashimoto called for a needed "push to enlarge the horizon of our foreign policy" beyond the Asia-Pacific, rediscovering a new Asian frontier, comprised in large part of the CA republics and the Caucasus.[45] Hashimoto hoped to integrate Japan and the CA states into a network of interdependence through the participation of the Japanese corporate community in resource exploration in this part of the world. For Hashimoto, the areas of interaction included, first and foremost, assisting these states in establishing affluent, prosperous domestic systems under a new political and economic structure. Japan also aimed to utilize the great potential of these states to serve as bridges to create distribution routes within the Eurasian region. In terms of its particular directions, Japan's Silk Road Diplomacy prioritized three main areas of concern: political dialogue; economic cooperation, including cooperation for natural resource development; and cooperation in peace-building, nuclear non-proliferation, democratization and fostering stability. The ODA assistance commitments of Japan in the CA region increased from US$2.57 million in 1993 to US$24.227 million in 2003, increasing ten times in ten years. Eventually, by the same year, the accumulated ODA disbursements on the bilateral level to the CA and Caucasus states reached US$1.98 billion.[46] The most notable among the successors of Hashimoto was PM Koizumi, who forwarded the most ambitious regional institution building in CA by first dispatching a "Silk Road Energy Mission" in July 2002 to promote closer energy cooperation with the CA states.[47] Emphasizing Japanese gains in terms of ODA and a positive attitude toward Japan, in March 2003, a group of Japanese experts concluded that CA is a new "frontier" in Asia where the Japanese presence can be further expanded.[48] In August 2004, Japan launched its CA plus Japan Dialogue forum.[49] PM Koizumi was the first Japanese PM to visit the CA region in 2006, traveling to the largest regional states of Uzbekistan and Kazakhstan.

At the same time, Japan's approach demonstrated its limitations and weaknesses. First, despite historical references to Silk Road connections, the Japanese leadership's attention often focused excessively on the ASEAN (Association of Southeast Asian Nations) countries while poorly integrating CA into overall Japanese foreign policy.[50] It is well reported that the CAJ dialogue scheme was built on a pattern of engagement between Japan and ASEAN in adapting "ASEAN Plus 3" to the CA context.[51] Some analysts at the Ministry of Foreign Affairs (MOFA) and in the expert community now suggest that this scheme can be used to create an ASEAN-like CA organization.[52] In addition, defining the role and importance of CA for Japan in practical terms has been difficult due to the distance and differences in social and legal systems along with the lack of connectivity with the Japanese corporate world. While this distance could be an asset for Japanese attempts to frame their CA engagement as "selfless" and a genuine contribution to CA development[53] such difficulty in defining the importance of CA for Japan also led to a lack of strategic planning and overall vision for Japanese engagement. As recognized by several diplomats who established Japanese embassies in the region, most of the initiatives were achieved spontaneously based on the enthusiasm and commitment of ambassadors and officials of the MOFA, Ministry of Finance and related agencies.[54] Japanese Ambassador Magosaki recalls that officials of the Ministry of Finance on a visit to Uzbekistan were touched by Uzbek Ministry of Finance officials' long work hours because they were reminded of work at the post-war Japanese ministries and the patriotism of Japanese civil servants in the immediate post–World War II years. Thus, Ambassador Magosaki recollects that, based on that sympathy, many decisions have been made to support Uzbek statehood construction.[55] Similarly, a few spontaneous attempts were made to adapt Japanese experiences of development to the CA region, for instance, by exporting the One Village One Product scheme of community empowerment that was successfully implemented in Japan in the 1970s. This scheme was introduced to Kyrgyzstan with varying degrees of success.[56]

China and Japan utilized the rhetoric of constructing partnerships with regional states. Although both emphasize the notion of partnership, there is a significant degree of difference when it comes to the practical implementation of these engagement schemes. For China, good-neighbor partnerships meant rather closed regional groupings in which China assumed the leading role. Japan employed open regional partnerships rhetoric rooted in the so-called "universal values" exemplified by democratic governance, a liberal market economy and transparency of the legal system, to name a few. The Japanese government also signaled the importance of the decolonizing nature of its initiative in this region by emphasizing the need for the initiative to come from CA countries in constructing regional cooperation under Japanese mediation. This emphasis was intended to empower regional states and strengthen their capability to address regional problems. In this Japanese scheme, the CA states were encouraged to seek alliances with each other while Japan would provide technical and financial assistance for the implementation of such initiatives. However, as many in the MOFA of Japan perceive, this expectation has been over-inflated, and there remains a gap in the understanding of the importance of cooperation among CA states. In contrast, no universal values are mentioned in the Chinese vision of region building; instead, the importance of respect for sovereignty and the "Shanghai spirit of cooperation" are emphasized.

The distinctive feature of the recent visit of PM Abe in 2015, detailed in the section below, is its prioritization of a more goal-oriented, practical approach to cooperation with CA, focused on functionality and the practical aspects of such cooperation over the value-based approach. Such a focus may be part of the shift in Japanese CA policy, within which the process of democratization is a longer-term objective while the economic opportunities of cooperation are the short-term goal. Although China has a longer history of such pragmatic cooperation with regional states, Japan has the advantage of a greater degree of trust (from both the general public, as seen from the data below, and the leadership) in this region. This trust paired with the new programmatic approach of Japanese leadership may yield greater results for Japan than in previous years.

Intensification of CA engagements: OBOR/BRI and the Japanese Abe mission of 2015

The Chinese OBOR/BRI initiative aims to expand the Chinese presence in this region and facilitate Chinese economic expansion beyond CA. In this sense, the Silk Road Economic Belt (consisting of six economic corridors, of which the Eurasian Land Bridge and China–CA–West Asia are relevant to CA, along with the "Twenty-First Century Maritime Silk Road") and the OBOR/BRI concept represent more sophisticated initiatives than SCO construction.[57] To provide stable financial support for the OBOR/BRI initiative, member states under Chinese leadership established the AIIB (in partnership with the Brazil, Russia, India, China and South Africa New Development Bank, the Silk Road Fund, the SCO interbank association, etc.) as an alternative to regional and global financial institutions such as the Asian Development Bank.[58] The infrastructure-related projects under this initiative also aim to address the lack of trust by some of

China's smaller neighbors by facilitating new initiatives for development infrastructure. These initiatives also aim to signal to CA states that China is genuinely interested in contributing to CA regional development.[59] This non-coercive, non-military (non-security focused) approach is supposed to further strengthen the soft-power potential of China in CA.

However, many in CA fear that, in practice, an OBOR/BRI initiative might become a facet of Chinese economic and cultural expansion.[60] Such fears caused an increase in Sinophobia in Tajikistan, Kyrgyzstan and Kazakhstan even before the OBOR/BRI plans were announced.[61] These sentiments have the potential to alienate the smaller members of the initiative.

In this sense, OBOR/BRI aims to address several issues through the same initiative: on the one hand, it helps shape a more positive image of China in CA, thus contributing to the construction of a sense of common belonging. On the other hand, Chinese engagement through OBOR/BRI is also part of China's response to the calls from regional states to create a common identity. To reflect a renewed mission to instill confidence toward China in CA, Chinese officials at various levels emphasize that OBOR/BRI is a mutually beneficial scheme, not an example of the "expansion of China".[62]

In terms of Japanese involvement, similar calls for a more mutually beneficial structure of relations were heard from CA countries during the visit of Japan's PM to CA in October 2015.[63] The visit of PM Abe to CA can be termed historic because it was the first time a Japanese PM had visited all five CA states. PM Koizumi visited CA in 2006, but his visit was limited to Uzbekistan and Kazakhstan. PM Abe's visit built upon previous Japanese engagement strategies exemplified by PM Hashimoto's Eurasian Diplomacy of 1997, the Obuchi Mission of 1998, FM Kawaguchi's visit of 2004 (when the CAJ initiative was announced) and PM Koizumi's visit of 2006, as mentioned above. Japanese experts generally attempt to differentiate the engagement of Japan in CA from that of China by arguing that Chinese "assistance" to CA is largely a gesture of goodwill similar to a "fireworks shoot" before the launch of major infrastructure-related projects.[64] For them, Japanese involvement in this region goes beyond infrastructure construction and attempts to transfer technology and knowledge.

The goals of PM Abe's visit to CA in 2015 partly confirm the description above and can be viewed as four-fold. First, the Japanese PM attempted to deepen and strengthen the presence of the Japanese business community in CA, as exemplified by the contracts signed during the visit for the joint exploration of gas fields in Turkmenistan (Galkynysh), Uzbekistan and Kazakhstan. This intensification of direct investments by Japanese companies with the support of the Japanese government has been encouraged by the majority of CA governments, as exemplified by the speech of the Tajik president, who explicitly emphasized the importance of switching from humanitarian aid projects to direct investment by Japanese companies.[65] Second, it was an attempt by the Japanese PM to secure orders from CA countries for Japanese corporations, as exemplified by the Japanese-Kazakh commitment to work on the construction of a nuclear plant in Kazakhstan and the Japanese-Turkmen agreement (between Turkmengas and

Japan Oil, Gas and Metals National Corporation) on the construction of mineral resource processing factories in Turkmenistan. Currently, Kazakhstan is also negotiating with Russia for the possible construction of the second such plant. Third, PM Abe aimed to further boost the "Cool Japan" soft-power construction initiative by supporting the construction of a Japanese university in Turkmenistan, cooperation in IT education in Tajikistan by launching the Youth Technological Innovation Center in Uzbekistan and similar educational initiatives in Kazakhstan.[66] Fourth, Abe's visit offered further humanitarian aid to the smaller republics of Tajikistan and Kyrgyzstan for various development-related projects.

Some experts have argued that Japanese technologies and loans for the projects mentioned above are a sign of competition between China and Japan because they offer alternative and thus competing sources of funding.[67] Some have even suggested that "incongruent interests between the two powers already hint for the potential for a friction in the region".[68] However, there is little evidence to suggest that these loans for Japanese projects in Turkmenistan are intended to affect Chinese projects. Although the field of mineral extraction coincides with Chinese interests, Japanese loans and projects in Turkmenistan do not aim to exclude China's investment there. In addition, the Japanese initiatives extend beyond energy resources to encompass human resource development, joint university and research facilities construction and human security infrastructure.[69] Additionally, former and current MOFA officials frequently attempt to define the features of "Japanese-ness" in assisting and engaging CA, which differ from the features offered by their Chinese counterparts.[70] Therefore, it is difficult to argue that Japan directly links its initiatives to the intensification of Chinese or Russian initiatives in this region. Its attempts to secure access to CA for Japanese corporate and state institutions may offer alternative resources for funding and development initiatives to those offered by China and Russia. However, at least in the official discourse and project implementation, Japanese involvement in CA does not appear to be an effort to counter Russian or Chinese initiatives because the goals and the degree of commitment to the region differ among the global players. Additionally, and interestingly, suggestions from the expert community in Japan indicate that OBOR/BRI initiatives in CA do not conflict with Japanese engagement. Rather, these experts suggest that Japan should utilize China-built and -financed infrastructure for the benefit of Japanese corporate penetration.[71] As has been noted by PM Abe, the OBOR/BRI has potential, and it is important that both Japan and China contribute to international peace, prosperity and the resolution of problems; therefore, Japan is prepared to cooperate (with China) when it can.[72] At the same time, PM Abe emphasized that OBOR/BRI infrastructure schemes should be transparent and that assistance and investments provided under this scheme should be repayable by the recipients without indebting these developing nations.[73] This comment echoes the views of the Japanese expert community, which favors refraining from challenging the Chinese financial and industrial presence with more targeted small-scale, but effective Japanese contributions to various projects.[74] The Chinese MOFA welcomed this Japanese stance by noting that OBOR/BRI will become a "development platform, creating benefits for countries around the world, including Japan" and responding to the Japanese PM's remarks with the statement that "China is committed to establishing a set of fair, reasonable and transparent rules for international trade and investment".[75] Therefore, both countries' foreign policies are obviously involved in an indirect process of mutual shaping and dialogue, which again confirms constructivist logic.

Challenges for Chinese and Japanese soft-power construction

Both China and Japan realize the importance of soft-power construction for achieving their goals in this region. Both countries utilize similar strategies, leading to talk of competition between them for the "hearts and minds" of the CA public. However, they also face challenges that are sometimes similar but more often different in nature.

Japan has been a forerunner in terms of establishing Japanese centers and language programs in this region. Japan's foreign policy traditionally emphasizes the importance of capacity building. One successful example of such a capacity-building institution is the case of the Japan Human Resource Development Centers in these countries. Their activities aim to provide Japanese language training and share Japanese expertise on business-skill development.[76]

As if to respond to these challenges and to support new institutional initiatives in this region, during the 2015 visit to CA, PM Abe announced in Tashkent that Japan and Uzbekistan would invest in launching centers that were not focused purely on culture and language but that had practical importance for development of the economy of CA countries and would generate a younger generation of "producers of knowledge". To exemplify this approach, PM Abe and his Uzbek counterpart announced the establishment of Uzbek–Japan youth innovation centers in partnership with the Japanese universities of Nagoya and Tsukuba. An agreement on the establishment of such centers was signed on October 26, 2015, in Tashkent. Although the structure of the centers and their activities have yet to be decided, the signing of the agreement represents both expectations that Japanese educational initiatives will be more widely applicable for Uzbeks and an attempt to more actively utilize the "friendly" social environment in Uzbekistan for wider penetration of the economic field by the Japanese corporate world. Additionally, values such as a shared mentality and respect for elders are also considered a good basis for cooperation in labor resource mobility between Uzbekistan and Japan. In particular, Japan, in cooperation with the Uzbek Ministry of Employment and Labor Relations, established a training center in Tashkent to provide pre-dispatch training to Uzbek personnel, who are to be sent to Japanese social security institutions (elderly care centers, etc.) for internships and time-limited work.[77] The intention of dispatching medical personnel from Uzbekistan to Japan is to utilize the excessive young labor resources available in Uzbekistan to fill the gap of resource deficiency in certain sectors of the Japanese labor market, such as social institutions for elderly people.[78] This program is also an attempt to take advantage of the dominant mentality in Uzbekistan of respect for the elderly, which can be practically utilized in Japan.

Chinese goals in soft-power construction resemble those of Japan, as described above. China is aware of and concerned about the importance of positive image construction in CA. However, the Confucian institutes and cultural/linguistic centers constructed in many republics of CA are often regarded with suspicion by many observers as comprising part of a strategy to culturally and linguistically facilitate the smoother expansion of Chinese economic interests.[79]

Thus, it is instrumental for China to improve its public image. Otherwise, any initiative China might launch and support will inevitably be linked to the hypothetical Chinese domination of this region. In terms of problems, the first issue China needs to confront is the image of Chinese economic expansionism. This image has been progressing in the minds of the public in Kyrgyzstan, Kazakhstan and Tajikistan throughout the 2000s. The second problem of Chinese soft-power creation in this region is an increasing discrepancy between the views of political leadership and the public regarding the Chinese presence. Political leadership frequently welcomes Chinese penetration of their societies because in many cases this implies support for current governments and leaders. Additionally, Chinese corporate penetration often implies increased wealth for many within the upper levels of power in these societies. However, this penetration will only produce the desired soft power if the Chinese engagement brings increased living standards to the public.

Conclusions

As has been argued above, CA has historically been prone to claims of rivalries among various powers. To a large extent, the current rhetoric of competition for domination in this region is the outcome of a neo-imperial vision of international relations. However, the areas in which Japan and China operate primarily reflect a process of reconstructing their foreign policy engagements in search of new identities of "self".

This chapter analyzed the Chinese and Japanese engagement in post-Soviet CA through the lenses of how they attempt to discursively internalize this region into their own internal and external policies, their imagery of this region and of "self" in relation to it, the norms and concepts according to which they justify their actions in CA (communication) and CA responses to these policies.

Although this chapter demonstrated that in the discourses of both countries, the notions of connectivity and infrastructure development and mineral resources and political stability feature prominently, there is little ground to claim that they desire exclusivity in certain sectors of the economy or cooperation with their CA counterparts. In contrast, each country seeks to define its place in such cooperation while calling on the other to contribute to its schemes of engagement, as exemplified by the Chinese OBOR/BRI initiative and the Japanese response to it. In this sense, the narrative of competition for regional domination prevalent in English-language, Russian and some CA media is unproven by any empirical evidence. In contrast, as is reflected in both Chinese and Japanese official discourses, many of the projects conducted by both China and Japan are structurally compatible, if not supplementary, and do not imply exclusivity of interest.

At the same time, these countries employ different discursive strategies to frame their approaches and goals. These discourses are their competing visions of CA development and their roles in such engagements. Thus, the different paradigms of reasoning for their CA engagements result in a "rivalry of discourses" for the "hearts and minds" of the CA population. As is seen in the responses of the CA public, the public expectations of these states are rather complicated. Chinese engagements generate both optimism and caution, while the Japanese presence currently encourages expectations with few practical results.

Notes

1 Y Wang, "Offensive for Defensive: The Belt and Road Initiative and China's New Grand Strategy", *The Pacific Review* 29, no. 3 (2016): 455–463, doi:10.1080/0951274 8.2016.1154690.
2 T Summers, "China's 'New Silk Roads': Sub-National Regions and Networks of Global Political Economy", *Third World Quarterly* 37 (2016): 1628–1643, doi:10.108 0/01436597.2016.1153415.
3 M Li, "From Look-West to Act-West: Xinjiang's Role in China–Central Asian Relations", *Journal of Contemporary China* 25 (2016): 515–528, doi:10.1080/10670564.2 015.1132753.
4 H Yu, "Motivation behind China's 'One Belt, One Road' Initiatives and Establishment of the Asian Infrastructure Investment Bank", *Journal of Contemporary China* 26 (2017): 353–368, doi:10.1080/10670564.2016.1245894.
5 M Ye, "China and Competing Cooperation in Asia-Pacific: TPP, RCEP, and the New Silk Road", *Asian Security* 11 (2015): 206–224, doi:10.1080/14799855.2015.1109509.
6 "Torukumenisustan gaz den Shien", *Sankei Shimbun*, September 7, 2015, www. sankei.com/economy/news/150907/ecn1509070003-n1.html.
7 Tony Tai-Ting Liu, "Undercurrents in the Silk Road: An Analysis of Sino-Japanese Strategic Competition in Central Asia", *Japanese Studies* 8 (March 2016): 157–171.
8 Nomura Research Institute (NRI), "Uzbekisutan kyowakoku no seizou gyou no shinkou to Nihon kigyou ni totte no no Jigyou no kikai" [Development of Industry of Uzbekistan and Opportunities for the Japanese Companies], *Public Management Review* 161 (2016): 1–6.
9 Timur Dadabaev, " 'Silk Road' as Foreign Policy Discourse: The Construction of Chinese, Japanese and Korean Engagement Strategies in Central Asia", *Journal of Eurasian Studies* 9, no. 1 (2018): 30–41, doi:10.1016/j.euras.2017.12.003.
10 This is also in line with J Townsend and A King, "Sino-Japanese Competition for Central Asian Energy: China's Game to Win", *The China and Eurasia Forum Quarterly* 5, no. 4 (2007): 23–47.
11 "Rea a-su: Kazahu to kyoudo kaihatsu" [Rare metals: Joint development with Kazakhstan], *Nikkei Shimbun*, May 2, 2012.
12 G Marat, "Japan, Turkmenistan Sign Deals Worth over $18 Bln in Chemicals, Energy", *Reuters*, October 23, 2015.
13 Timur Dadabaev, "The Chinese Economic Pivot in Central Asia and Its Implications for the Post-Karimov Re-emergence of Uzbekistan", *Asian Survey* 58, no. 4 (2018): 747–769, doi:10.1525/as.2018.58.4.747.

14 M Rakhimov, "Central Asia in the Context of Western and Russian Interests", *L'Europe en Formation* 2015/1, no. 375 (2015) 140–154; EB Rumer, D Trenin and H Zhao, *Central Asia: Views from Washington, Moscow, and Beijing* (Oxon: Routledge, 2007).

15 F Tolipov, *Bolshaia Strategiia Uzbekistana v Usloviiah Geopoliticheskoi I Ideologicheskoi Transformatsii Tsentral'noi Azii* [Great strategy of Uzbekistan in the conditions of geopolitical and ideological transformation in Central Asia] (Tashkent: Fan, 2005).

16 S Blank and AZ Rubinstein, *Imperial Decline: Russia's Changing Role in Asia* (Durham, NC: Duke University Press, 1997).

17 A Cooley, *Great Games, Local Rules: The New Great Power Contest in Central Asia* (Oxford: Oxford University Press, 2012).

18 Nargis Kassenova, "China's Silk Road and Kazakhstan's Bright Path: Linking Dreams of Prosperity", *Ponars Eurasia*, 2017, www.ponarseurasia.org/article/china's-silk-road-and-kazakhstan's-bright-path-linking-dreams-prosperity, accessed October 2, 2017.

19 D Scott, "The Great Power 'Great Game' between India and China: 'The Logic of Geography'", *Geopolitics* 13 (2008): 1–26, doi:10.1080/14650040701783243; M Laruelle, S Peyrouse and B Balci, eds., *China and India in Central Asia: A New "Great Game"?* (New York: Palgrave Macmillan, 2010).

20 C Wallander, "Silk Road, Great Game or Soft Underbelly? The New US–Russia Relationship and Implications for Eurasia", Southeast European & Black Sea Studies 3 (2003): 92–104, doi:10.1080/14683850412331321668.

21 L Jonson, *Vladimir Putin and Central Asia: Shaping of Russian Foreign Policy* (London: IB Tauris, 2006).

22 Timur Dadabaev, "Japan's Search for Its Central Asian Policy: between Idealism and Pragmatism", *Asian Survey* 53 (2013): 506–532, doi:10.1525/as.2013.53.3.506; Timur Dadabaev, "Chinese and Japanese Foreign Policies towards Central Asia from a Comparative Perspective", *The Pacific Review* 27, no. 1 (2014): 123–145, doi:10.1080/09512748.2013.870223; Timur Dadabaev, "Shanghai Cooperation Organization (SCO) Regional Identity Formation from the Perspective of the Central Asia States", *Journal of Contemporary China* 23, no. 85 (2014): 102–118, doi:10.1080/10670564.2013.809982.

23 N Swanström, "China and Central Asia: A New Great Game or Traditional Vassal Relations?", *Journal of Contemporary China* 14, no. 45 (2005): 569–584, doi:10.1080/10670560500205001.

24 Y Kim and S Blank, "Same Bed, Different Dreams: China's 'Peaceful Rise' and Sino-Russian Rivalry in Central Asia", *Journal of Contemporary China* 22 (2013): 773–790, doi:10.1080/10670564.2013.782126.

25 M Laruelle and E McGlinchey, *Renewing EU and US Soft Power in Central Asia*, EUCAM Commentaries 28 (2017), www.eucentralasia.eu/uploads/tx_icticontent/EUCAM_Commentary_28_Renewing_US_and_EU_Softpower_in_Central_Asia.pdf.

26 D Trenin, From Greater Europe to Greater Asia: the Sino-Russian Entente (Moscow: Carnegie Moscow Center, 2015); A Pikalov, "Uzbekistan between the Great Powers: A Balancing Act or a Multi-Vectorial Approach?", *Central Asian Survey* 33 (2014): 297–311, doi:10.1080/02634937.2014.930580.

27 S Zhao, "Rethinking the Chinese World Order: The Imperial Cycle and the Rise of China", *Journal of Contemporary China* 24 (2015): 961–982, doi:10.1080/10670564.2015.1030913.

28 S Peyrouse and G Raballand, "Central Asia: The New Silk Road Initiative's Questionable Economic Rationality", *Eurasian Geography & Economics* 56 (2015): 405–420, doi:10.1080/15387216.2015.1114424.

29 On Japan–China balancing, see Kei Koga, "The Rise of China and Japan's Balancing Strategy: Critical Junctures and Policy Shifts in the 2010s", *Journal of Contemporary China* 25, no. 101 (2016): 777–791.

30 S Hakamada, "ChuRo wo Keikaisuru Chuouajia Shyokokuno Nihon he kitai wa okii" [Expectations are high in respect to Japan as a warning to China and Russia], *Sankei Shimbun*, October 23, 2015, www.sankei.com/column/news/151023/clm1510230001-n1.html.

31 On the new Great Game, see A Avagyan, "Novaya Bol'shaya Igra: Komu dostanetsya Tsenral'naya Aziya" [New great game: Who will get the Central Asia], Iarex.ru, September 23, 2015, www.iarex.ru/articles/52022.html. On Japanese revenge, see "Yaponiya nanosit otvetnyi udar" [Japan strikes back], *Liter*, October 30, 2015, www. liter.kz. On the expansion of Japanese influence, see "Yaponiya hochet ukrepit' svoyo vliyanie v Tazjikistane I TsA" [Japan wants to enhance its influence in Tajikistan and CA], *Sputnik*, October 3, 2016, www.sputnik-tj.com. And on resource diplomacy, see "Tsentral'noaziatskoe turne Sindzo Abe: Resoursnaya diplomatiya ili partiya v Bol'shoi Igre" [Central Asian tour of Shinzo Abe: Resource diplomacy or a play in a "Great Game"]? *Narodnyi Korrespondent*, November 26, 2015, www.hk.org.ua.

32 DA Mileev, "Vazaimotonosheniya Yaponii I SHOS" [Relations between Japan and SCO], Center for Research of Problems of Orient of Academy of Sciences of Russia (undated), www.vostokoved.ru/%D0%A1%D1%82%D0%B0%D1%82%D1%8C%D 0%B8/japshos.html.

33 S Bitkovski, "Yapono-Kitaijskoe Sopernichestvo v Stranah 'Tret'ego Mira'" [The rivalry of Japan and China in the 'Third World' countries], *Journal of International Law & International Affairs* 4 (2008): 64–69.

34 Townsend and King, "Sino-Japanese Competition for Central Asian Energy"; C Len, T Uyama and T Hirose, eds., *Japan's Silk Road Diplomacy: Paving the Road Ahead* (Washington, DC: Central Asia-Caucasus Institute and Silk Road Studies Program, 2008); T Uyama, "Chuyou Ajia Shyokoku kara mita kokusai kankyou no henka to taiou: Roshia no seijiteki/gunjiteki eikyouryoku to Chugoku no keizai shinshitsu" [Change and adaptation to the international environment from the perspective of CA states: Expansion of Russian political and military influence and Chinese economic penetration], *Kokusai Mondai* 647 (2015), pp. 24–25.

35 On the dynamics of Japan–China relations, see R Drifte, W Vosse and V Blechinger-Talcott, *Governing Insecurity in Japan: The Domestic Discourse and Policy Response* (Sheffield/London: Sheffield Centre for Japanese Studies/Routledge Series, 2003).

36 For details of cooperation road maps, see Timur Dadabaev, *Chinese, Japanese and Korean In-Roads into Central Asia: Comparative Analysis of the Economic Cooperation Road Maps for Uzbekistan*, Policy Studies Series (Honolulu/Washington: East West Centre, 2019, accepted, forthcoming). Dadabaev, "The Chinese Economic Pivot".

37 Dadabaev, "Chinese and Japanese Foreign Policies"; Dadabaev, "'Silk Road' as Foreign Policy Discourse"; Timur Dadabaev, "Engagement and Contestation: The Entangled Imagery of the Silk Road", *Cambridge Journal of Eurasian Studies* 2018, no. 2, doi:10.22261/cjes.q4giv6.

38 Dadabaev, "'Silk Road' as Foreign Policy Discourse".

39 Interview with FM Fumio Kishida, "Shinrai to sougo kanshin no kiban no ueni" [Building on the foundations of trust and mutual interest], *Neutral Turkmenistan*, May 9, 2017, in Japanese and Russian at www.mofa.go.jp/mofaj/p_pd/ip/page4_002989.html.

40 For "win–win" principles, see Rosita Dellios, "Silk Roads of the Twenty-First Century: The Cultural Dimension", *Asia & the Pacific Policy Studies* 4, no. 2 (2017): 225–236, doi:10.1002/app.5.172.

41 S Cornelissen and I Taylor, "The Political Economy of Chinese and Japanese Linkages with Africa: A Comparative Perspective", *The Pacific Review* 13, no. 4 (2000): 615–633, especially p. 617.

42 Dadabaev, "'Silk Road' as Foreign Policy Discourse".

43 Zaidan Hojin Kokusai Koryu Senta [Japan Center for International Exchange], ed., *Roshia Chuo Ajia taiwa misshon hokoku: Yurasia gaiko he no josho* [Report of the

Mission for Dialogue with Russia and Central Asia: Introduction toward Eurasian Diplomacy] (Tokyo: Roshia Chuo Ajia taiwa misshon, 1998).

44 Takeshi Yuasa, "Japan's Multilateral Approach toward Central Asia", in *Eager Eyes Fixed on Eurasia: Russia and Its Neighbors in Crisis*, ed. Akihiro Iwashita (Sapporo: Hokkaido University Slavic Research Center, 2007), www.src-h.slav.hokudai.ac.jp/coe21/publish/no16_1_ses/04_yuasa.pdf., accessed September 8, 2018.

45 R Hashimoto, Address to the Japan Association of Corporate Executives, Tokyo, July 24, 1997, www. Japan.kantei.go.jp/0731douyuukai.html.

46 Ministry of Foreign Affairs (MOFA) of Japan, Department for International Cooperation, ed., *Seifu kaihatsu enjo (ODA) kunibetsu deta bukku, 2004* [Japan's Official Development ment Assistance: Annual Report, 2004] (Tokyo: MOFA, 2005), 199–204.

47 MOFA of Japan, *Shirukurodo Enerugi mishon* [Silk Road energy mission]. (Tokyo: MOFA, 2002).

48 Japan Institute for International Affairs (JIIA), *Chuou Ajia ni kansuru teigen* [Recommendations regarding Central Asia] (Tokyo: JIIA, March 2003), www2.jiia.or.jp/pdf/russia_centre/h14_c-asia/03_kasai.pdf.

49 T Yagi, "'Central Asia plus Japan' Dialogue and Japan's Policy toward Central Asia", *Asia-Europe Journal* 5 (2007): 13–16.

50 Dadabaev, "Japan's Search for Its Central Asian Policy"; Dadabaev, "Chinese and Japanese Foreign Policies".

51 A Kawato, "What Is Japan Up To in Central Asia?", in *Japan's Silk Road Diplomacy: Paving the Road Ahead*, ed. C Len, T Uyama and T Hirose (Washington, DC: Central Asia-Caucasus Institute and Silk Road Studies Program, 2008), p. 16.

52 "Chuou Ajia hatten no kokusaiteki jyoken to Nihon" [Development of Central Asia: International conditions and Japan], *Gaiko* 34 (2015): 21–34, especially pp. 32–33.

53 Len, Uyama and Hirose, eds. *Japan's Silk Road Diplomacy*, p. 111.

54 N Murashkin, "Japanese Involvement in Central Asia: An Early Inter-Asian Post-Neoliberal Case?" *Asian Journal of Social Science* 43 (2015): 50–79.

55 Ambassador Magosaki's recollection of his speech at Keio University, Tokyo, http://web.sfc.keio.ac.jp/~kgw/watergovernance/Magosaki.pdf.

56 Timur Dadabaev, "One Village – One Product: The Case of JICA's Community Empowerment Project in Kyrgyzstan", in *Japan in Central Asia: Strategies, Initiatives, and Neighboring Powers*, ed. T Dadabaev (New York: Palgrave Macmillan, 2016), pp. 69–85.

57 Office of the Leading Group for the Belt and Road Initiative, *Building the Belt and Road: Concept, Practice and China's Contribution* (Beijing: Foreign Languages Press, May 2017), pp. 11–17, https://eng.yidaiyilu.gov.cn/wcm.files/upload/CMSydy-lyw/201705/201705110537027.pdf, accessed March 12, 2018.

58 National Development and Reform Commission, Ministry of Foreign Affairs and Ministry of Commerce of People's Republic of China, "Vision and Actions on Jointly Building Silk Road Economic Belt and 21st-Century Maritime Silk Road", March 30, 2015.

59 For instance, see Ministry of Foreign Affairs of the People's Republic of China, "Joint Declaration on New Stage of Comprehensive Strategic Partnership Between the People's Republic of China and Republic of Kazakhstan", August 31, 2015, www.fmprc.gov.cn/mfa_eng/wjdt_665385/2649_665393/t1293114.shtml, last seen on January 25, 2018).

60 M Auezov, "Ex-Ambassador of Kazakhstan to China Concerned over China's Classified Documents", *Tengri News*, 2015, www.en.tengrinews.kz/politics_sub/Ex-Ambassador-of-Kazakhstan-to-China-concerned-over-Chinas-21645/, last accessed on January 15, 2016.

61 R Khodzhiev, "V svoyom dome ne khoziaeva: Kitaiskaia ekspasiya v Tazhikistan", *Centrasia.ru*, November 1, 2011, www.centrasia.ru/news.php?st=1320094800 (last accessed on November 1, 2011).

62 "Posol: proekt 'Odin poyas, odin put' – eto ne ekspansiya Kitaya" [Ambassador: "One Belt, One Road" – is not expansion of China], *Podrobno.uz*, May 29, 2017, www.podrobno.uz, last accessed on May 29, 2017.
63 "Atambaev and Sariev Meet with Shinzo Abe", *Vestnik Kavkaza*, October 25, 2015, http://vestnikkavkaza.net/news/Atambaev-and-Sariev-meet-with-Shinzo-Abe.html.
64 "Chuou Ajia hatten no kokusaiteki jyoken to Nihon".
65 Joint Statement of Tajik President and Japanese PM, 2015, Ministry of Foreign Affairs, Japan.
66 Joint Statements of Japanese PM and CA Presidents, 2015, Ministry of Foreign Affairs, Japan.
67 M Rakhimov, "Central Asia and Japan: Bilateral and Multilateral Relations", *Journal of Eurasian Studies* 5, no. 1 (2014): 77–87, doi:10.1016/j.euras.2013.09.002.
68 Liu, "Undercurrents in the Silk Road".
69 Interview with FM Fumio Kishida.
70 "Chuou Ajia hatten no kokusaiteki jyoken to Nihon".
71 "Chuou Ajia hatten no kokusaiteki jyoken to Nihon".
72 "Shyusho 'Ittai ichirou ni kyoryoku' Hatsuno hyoumei" [PM first announcement of preparedness for cooperation along "OBOR/BRI"], *Yomiuri Shimbun*, June 5, 2017.
73 "Chugoku no 'Ittai Ichiro' Kyoryoku ni Seimeisei, Kouseiseinadoga 'jyouken'" [Transparency and legitimacy are the conditions for support of Chinese OBOR], *Sankei Shimbun*, June 5, 2017, www.sankei.com, last seen on June 8, 2017.
74 H Endo, "Abe Shyusho Chuou Ajia Rekiho to Chugoku no Ittai Ichiro" [PM Abe's CA visit and Chinese OBOR], *Yahoo!*, October 26, 2015, https://news.yahoo.co.jp/byline/endohomare/20151026-00050809/.
75 "Foreign Ministry Spokesperson Hua Chunying's Regular Press Conference on June 6, 2017", www.fmprc.gov.cn/mfa_eng/xwfw_665399/s2510_665401/t1468248.shtml.
76 Dadabaev, "One Village – One Product".
77 Embassy of Japan in Uzbekistan, "Tseremoniya otkrytiya 'Tsentra po predvyezdnoi adaptatsii I obucheniyu tehnicheskih stazherov'" [Ceremony of launching pre-dispatch adaptation and training of technical interns], September 14, 2017, www.uz.emb-japan.go.jp/itpr_ru/00_000122.html.
78 "Uzbekistan to Supply Medical Personnel to Japan Labor Market", *The Tashkent Times*, May 15, 2017, http://tashkenttimes.uz/national/929-uzbekistan-to-supply-medical-personnel-to-japan-labor-market.
79 Dadabaev, "Shanghai Cooperation Organization (SCO) Regional Identity Formation".

4 Japanese and Chinese infrastructure development strategies in Central Asia

In the years following the collapse of the Soviet Union, various infrastructure facilities in Central Asia (CA) remained as legacies of Soviet policies. The majority of railroads have historically connected CA to Europe and the Middle East by passing through Russian territory. The goods transported through energy transportation networks were similar to those of the Soviet era, which included mostly mineral resources (oil, gas, etc.). Agricultural products (such as cotton) were also transported to Russian and other East European markets. Therefore, many have claimed that the infrastructure constructed in the CA region during the Soviet years, although very beneficial for regional states, still constitutes part of the Soviet and post-Soviet colonial structure as far as CA is concerned. In addition, the products generated by CA producers frequently duplicate each other, thus making CA states competitors on the international market and demotivating the development of infrastructure between CA states. Therefore, the rise of China and the increasing role of Japan in this region have both been cautiously welcomed by the expert community and politicians because they are bringing new infrastructure projects of a decolonizing nature and not only linking China and Japan to CA states but also linking regional states to each other.[1] Some of these new initiatives have included the China-supported BRI (Belt and Road Initiative) infrastructure projects and the creation of the Asian Infrastructure Investment Bank (AIIB). Japan-supported infrastructure (related to ODA, Central Asia Regional Economic Cooperation (CAREC), the Quality Infrastructure concept, etc.) in the CA region, financed through the Asian Development Bank and the Japan Bank for International Cooperation, has also been hailed as supporting the independence of these countries and providing necessary networks to sustain and develop their economies. At the same time, such intensification of infrastructure-related projects in the CA region has led many to describe this situation in alarmist tones.[2] The most recent example of such alarmist rhetoric is anti-Chinese riots in Kyrgyzstan where local residents protested Chinese investments in gold mining as something which brings destruction to local livelihoods and environment.[3] Prime Minister (PM) Iskakov announced that these protests were fueled and supported by certain politicians and political forces. Such riots and protests are not exceptional and have happened in CA frequently in recent years.[4] In particular, certain warning concerns have been voiced about the neo-colonizing potential of Chinese-financed and

Chinese-constructed projects and about the possibility of a new Great Game among various powers – including Japan – as part of the rivalry over regional resources and infrastructure projects.[5] The motivations behind these powers' involvement in infrastructure-related projects have been questioned.

In response to such alarmist rhetoric, this chapter enquires into the motivations behind China's and Japan's engagement in infrastructure-related projects in the CA region. The main questions to be considered in this chapter are the following: How can Japanese and Chinese infrastructure development projects in the CA region be interpreted and narrated? What are the similarities and differences in how these two countries frame their approaches to infrastructure development?

To answer these questions, this chapter will analyze the motivations of Japan and China through the following layers of analysis. The chapter will first provide a concise overview of the general foreign policy orientations of Japan and China in CA to illustrate the framing of infrastructure development strategies in this region. It will then consider the relevance of Japanese and Chinese infrastructure to the notions of human and regional security and subsequently consider the relevance of the economic aspects of these infrastructure projects to the roles played by China (biggest trade partner) and Japan (one of the largest aid providers). The chapter will then focus on two particular projects, namely, rail and energy infrastructure–related projects conducted by China and Japan, to highlight the differences in the approaches of these two states. Due to word limitations of this chapter, it would be impossible to cover all the CA countries. Thus, the two cases of railroad infrastructure are selected from a centrally positioned country in CA, namely Uzbekistan, to illustrate the argument.

This chapter develops several arguments with respect to the Japanese and Chinese approaches to infrastructure development in CA. First, in line with criticism of the "new Great Game", sentiments voiced inside and outside of the CA region,[6] this chapter argues that the discourse of mutually exclusive interests in the development of various infrastructure-related projects in CA by China and Japan is premature and largely unproven. Most of the Chinese engagements emphasize energy and transportation infrastructure creation (construction), while for Japan, the main fields of focus are the areas of current infrastructure maintenance, modernization and rehabilitation. Thus, this chapter suggests that China positions itself as the largest economic partner to CA, while Japan is the largest assistance provider. These two roles have different implications. Second, the current infrastructure engagements of Japan (from assistance to partnership) and China (from exploitation to contribution to the region) in CA demonstrate the attempts at adjustment motivated by both countries' search for new identity formation and standing in the region.

Foreign policy orientations with respect to CA and infrastructure development

Any narration of infrastructure development in CA by China and Japan needs to be considered by integrating the two countries' infrastructure construction within their general foreign policy engagements in the CA region. As has been discussed in the literature, the collapse of the Soviet Union presented both China and Japan with a new frontier of foreign policy engagement.[7] As is described in Chapters 2 and 3, while for Japanese foreign policy this new frontier required starting from scratch, for the Chinese the launch of newly established relations has been further complicated by the number of problems left unresolved from the era of the Soviet Union, the most important of which were land/border claims. Therefore, it was natural for the Chinese government to attempt to first strengthen trust with CA regional states and decrease the level of tensions in the bordering areas through "Four plus One (China plus Russia, Kazakhstan, Kyrgyzstan and Tajikistan)" and – later in Shanghai – five negotiation processes.[8] In this process of constructing "good-neighboring relations", the general principles of the Chinese foreign policy of non-interference in internal affairs and the seeking of mutual benefit and common prosperity were well received by many of China's CA counterparts, contrasting with the constant criticism of CA states' domestic policies by the Western states in Europe and North America.[9] In addition, the Chinese attitude also signified Chinese resentment of Western criticism of its domestic human rights record. Thus, both China and the CA states have come to realize that they have not only shared concerns but also shared values (Shanghai spirit) and shared international constraints (such as criticism of their governance styles and fear of interference in internal affairs, to name a few), which eventually shaped Sino-CA relations.[10] With the formation of the Shanghai Cooperation Organization, successes in security-related issues spilled over into the area of economic cooperation, paving the way for a number of infrastructure development projects that eventually peaked with the announcement of the Silk Road (BRI) initiative and the goal of improving the connectivity of the CA region both with China and with other countries.[11]

Therefore, the infrastructure development projects conducted in the CA region are part of the Chinese policy of developing "good-neighbor" relations with this region. Such "good-neighbor" rhetoric includes promoting the interconnectedness of China and its CA counterparts to meet the economic development needs of both CA regional states and Chinese bordering provinces. In the discourse of the "good neighborhood", establishing a secure neighborhood for China is also connected to infrastructure projects, which induce economic development.[12] Economic development, in turn, is considered to be a pillar of sustainable stability and security.[13] Building a "good neighborhood" in CA was rooted in the initiatives established between China and its CA counterparts long before BRI was announced and is connected to the border delimitation and confidence-building initiatives of the early 1990s, which later led to the Shanghai Cooperation Organization (SCO) and eventually incorporated the CA region into BRI. Additionally, the Shanghai spirit, which implies common decision making and common benefits, is similarly reflected in the discursive "selling" of the BRI initiative to CA states. Interestingly, Chinese multilateral initiatives (e.g., BRI) are being simultaneously developed with bilateral initiatives, such as strategic partnership agreements with Kazakhstan, Tajikistan and Uzbekistan.[14] These strategic partnerships often refer to multilateral agreements, and it is often difficult to distinguish which of the initiatives are purely bilateral and which are of a multilateral nature. For the Chinese expert community, both are consistent parts of Chinese engagements, with each one supporting the other.

For the Chinese government, bringing infrastructure to CA is not "charity" but rather part of the realization that connectivity and infrastructure development can be additional building blocks in constructing a new international identity for China, in which the country not only exports goods and services to other countries but also constructs an area in which Chinese approaches (to development) and values (governance and inter-state relations) are accepted, shared and further developed. The "Silk Road" narrative strategy has served to further a convenient and historically based discourse that is easily understood and accepted by CA counterparts because it "paints" the CA region as arguably central to the success of the whole project.

For the Japanese government, the launching of its engagement with the CA region was less complicated than the Chinese engagement in logistical terms. Japan did not have any unresolved problems or issues with CA republics. In contrast to East Asia, there are no images of Japanese imperialism or neo-colonialism in CA. In addition, Japan projected the image of being both an economic superpower and the second largest economy in the world. Furthermore, as is explained in Chapter 3, the Japanese government and its officers at the Ministry of Foreign Affairs and Ministry of Finance often expressed sympathy toward colleagues in newly established partner ministries.

By the mid-1990s, the Japanese government had come to recognize a new international environment in which Japan's standing needed to be sustained, and Eurasia would become the next new frontier. This was the theme of PM Hashimoto's address in 1997[15] preceded by the Obuchi Mission of 1997. However, it was only during the administration of PM Koizumi that the Japanese PM first visited CA, launching the institution building process between CA and Japan.

At the same time, the largest difference between the Chinese and Japanese engagements in this region is that Japan, in its infrastructure development and aid allocation, emphasizes universal values such as democratic governance, transparent procedures and human rights, as well as Asian values of cooperation and step-by-step progress.[16] In addition, Japan is part of the OECD (Organisation for Economic Co-operation and Development) and its DAC (Development Assistance Committee) which also sets certain standards regarding provision of developmental assistance. In practical terms, this means that Japan is willing to support many infrastructure development projects in CA as long as the transparency of financial flows and the technical implementation of projects are guaranteed. This is not to say that the Japanese government and corporations can completely avoid corruption and the deficiency of governance procedures. In contrast, on certain occasions, the Japanese projects and a number of government officials have been pressured to provide irregular payments to local officials to proceed with certain initiatives (particularly in Uzbekistan; see Chapter 7 for details). However, such situations, when uncovered by the

Japanese government and press, have been denounced, and the aid to CA countries involved in such schemes has been reduced.[17] The Japanese ODA charter also clearly emphasizes the importance of transparency and good governance (including eradication of corruption) as necessary conditions for Japanese involvement in infrastructure development and assistance. This hybridity of Japan's value orientations in its relations with CA states demonstrates the duality of its value orientations both domestically and internationally. While Japan displays some understanding of various problems facing CA states, it also emphasizes some Western values. At the same time, Japan is also often accused of violations of these values, which the Japanese government rebuffs as Japan-bashing, similar to the CA rebuffing of international criticism.

Framing of infrastructure projects

In their infrastructure development projects, both China and Japan attempt to use frames that are easily accepted by the host countries and international community. However, their branding of their infrastructure development – and their selling points for such projects – differ significantly, reflecting the different national and international standing of each country. In 2013, on a visit to CA, the Chinese president announced that the launch of the Silk Road Economic Belt had five different implications.[18] For the Chinese government, the initiative is first a foreign policy component of the realization of the "Chinese dream" strategy announced by the president of China. Second, it represents "bridging" between Chinese producers and international markets through two sets of roads (continental and maritime). Third, as explained below, it is an attempt to establish a stable, prosperous and thus secure neighborhood by revitalizing China's presence in nearby areas.[19] Fourth, it is also a strategy to bring about the development of various provinces in China by increasing demand for their products. Finally, it contributes to increasing interdependence between China and other countries through BRI and thus to strengthening their mutual relations. The CA region is considered to be a key region in the construction of one such Silk Road – namely, the Eurasian Land Bridge. As analyzed in Chapter 2 of this study, the notion of the Silk Road is an easily understood branding strategy, which is assisted by its historical connotations. It is also a concept that is open to interpretation and places different states at its center. To the Chinese, China is the launch pad and the main source of both financing (through the AIIB and other instruments) and ideas behind this concept. For CA states, however, the narrative of this initiative puts the CA region at center stage of the project because, presumably, without the participation of CA states, there will not be a Silk Road. Additionally, CA states consider this road to be a road for job creation and technology transfer as opposed to a simple Chinese trade route. Despite such differences in "readings", the Chinese BRI initiative deliberately maintains the validity of all types of interpretations to motivate member states and facilitate the smooth construction of the relevant railroad and energy resource infrastructure.

There are three main components of the Chinese BRI scheme that relate to CA states. First, all previous projects related to the import of gas and oil from CA to China (for instance, the 2,800-km pipeline from Kazakhstan to China of 2006 and the Sino-Kazakh logistics hub of 2014) are to be integrated into the BRI initiative as a part of a grand strategy. As discussed below, there were many strategic agreements between China and its CA counterparts and energy resource–related projects that were developed on a project-based level prior to the BRI announcement. However, with the announcement of the BRI initiatives, there was an attempt to integrate and "fill in" the BRI initiative with as much substance as possible to emphasize the diversity of its directions and its importance to CA.

Second, Chinese infrastructure-related projects are not an attempt at development mentorship. These infrastructure facilities do offer a business model or a model of development (in terms of certain technologies, know-how and the way the infrastructures can be utilized), but they are not an attempt to impose a "Chinese way of doing things". In contrast, these projects constitute a pattern of economic cooperation in which the Chinese government and participants ensure that their interests are taken care of, while it is up to the CA counterparts to make sure that the arrangement serves their national interests in a "win–win" manner.[20] It is often the case that CA states do not have the capacity or political will (because of corruption) to ensure that their national interests are properly guaranteed in infrastructure projects, and they blame the Chinese side for mishaps. However, the general Chinese approach to such projects has a pragmatic "partnership not mentorship" logic.

Third, the notion of the mutual complementarity of goals behind infrastructure-related projects is another backbone of Chinese engagements, in which Chinese projects often attempt to link and complement national development goals. In this way, the Chinese government aims to ensure that infrastructure development not only serves Chinese producers and consumers but also receives commitment from their CA.[21] The outcomes do not necessarily correspond to such declared goals. Frequently, incorrectly identified national goals might lead to problems with Chinese infrastructure development. In terms of infrastructure project implementation, the claims often made by CA governments and non-governmental organizations are that the poor quality and the environmental, demographic and social costs of Chinese projects outweigh their benefits.[22] In particular, the prime complaints of CA partners are that the Chinese corporations involved in the infrastructure development projects brought in their own workforces, although the necessary skills were often available in CA countries. In addition, the treatment of the local workforce has often been cited as discriminatory. Despite such obvious problems, discursively, the notion of mutual benefit has always been articulated in framing infrastructure development.

Japanese engagement in and support for infrastructure development has been driven by the following main principles. First, Japan, being distanced from the region and not having any common borders with it, favors the concept of "open regionalism". This implies that Japan, while attempting to build and enrich its CA plus Japan dialog, does not aim to build an exclusive regional forum, thus not counterposing it to any party wanting to play a constructive role in the region. Neither does Japan claim exclusivity in terms of participants. Such an approach corresponds to Japan's overall foreign policy engagements elsewhere and reflects its geographical proximity to the CA region. Another important factor that has forced Japan to adopt such a policy is that countries located between Japan and CA (China, Russia and South Korea) have complicated "love–hate" relations with Japan. In these countries, it is important to ensure the proper understanding and acceptance of Japanese engagement in CA to project a powerful image of Japanese relations with respect to the interests of these states. The notion of open regionalism suits such accommodation goals well.

Second, beginning with their independence, Japan has always played the role of a major donor to the CA states. Japanese officials and foreign policy practitioners have displayed paternalistic support toward the newly independent states of CA since the early years of their independence. Some claim – arguably – that this turned into a "mentorship" at times,[23] which, if true, contrasts with the Chinese approach of "partnership not mentorship" mentioned above. This role for Japan has led to commitments and support being channeled through the Ministry of Foreign Affairs and Ministry of Finance to Uzbekistan, Kyrgyzstan and Tajikistan. Uzbekistan, in particular, has enjoyed Japanese developmental aid support throughout its independence, which has resulted in Uzbekistan being the leading recipient of Japanese developmental aid in this region.[24] Additionally, Uzbekistan has shown a very high level of support for Japanese initiatives at both the political and public levels. Japanese priorities in terms of ODA allocation, however, have changed several times since the 1990s. At first, the logic of Japan's Ministry of Foreign Affairs was that smaller amounts of Japanese loans would be more efficient when extended to demographically and territorially smaller countries such as Kyrgyzstan and Tajikistan. However, with time, it became obvious that such loan disbursement was not very successful for several reasons, including weak state administration in Kyrgyzstan and Tajikistan and a lack of capacity to effectively use and eventually repay the Japanese ODA loans. This situation has forced a re-adjustment in the attitude of the Japanese government toward Kazakhstan and Uzbekistan. However, Kazakhstan displayed a desire to attract more direct Japanese investments as opposed to governmental ODA loans. Thus, the total disbursements of Japanese assistance, as it appears in the data outlined in Chapter 7 and in Table 4.1, tends to favor Uzbekistan.

Also, as seen in the flows of aid and direct investments from Japan to CA, the majority of infrastructure-related projects supported by Japan rely on ODA assistance. This results in a conceptualization of Japanese infrastructure development as "assistance, not economic partnership". This is in stark contrast to the Chinese approach of "economic partnership (not assistance)" outlined above.

To compensate for the shortcomings of the Japanese infrastructure-related engagement and to provide incentives for Japanese corporations to engage their Asian (including CA) counterparts, the Japanese PM has announced a

Table 4.1 Japan's international cooperation policy in Central Asia and the Caucasus region (calendar year of 2016, in US$ millions)

Rank	Country or region	Grants				Loan aid			Total (Net disbursement)	Total (Gross disbursement)
		Grant aid	Grants provided through multilateral institutions	Technical cooperation	Total	Amount disbursed (A)	Amount recovered (B)	(A)−(B)		
1	Uzbekistan	9.77	2.42	6.84	16.61	178.49	28.59	149.90	166.51	195.10
2	Azerbaijan	0.46	–	0.92	1.38	57.25	20.44	36.81	38.18	58.63
3	Tajikistan	25.41	8.27	5.63	31.04	–	–	–	31.04	31.04
4	Georgia	1.07	–	0.84	1.92	16.13	2.63	13.50	15.42	18.05
5	Kyrgyz Republic	4.48	–	8.33	12.81	–	0.48	−0.48	12.34	12.81
6	Armenia	2.02	–	2.62	4.64	–	10.32	−10.32	−5.67	4.64
7	Kazakhstan	0.31	–	1.99	2.30	0.66	38.84	−38.18	−35.88	2.96
8	Turkmenistan	–	–	0.50	0.50	–	2.01	−2.01	−1.51	0.50
	Multiple countries in Central Asia and the Caucasus	2.72	–	0.62	3.34	–	–	–	3.34	3.34
	Central Asia and the Caucasus region total	46.25	10.69	28.30	74.54	252.53	103.31	149.22	223.77	327.07

Source: Reproduced from MoFA of Japan site: www.mofa.go.jp/policy/oda/white/2017/html/honbun/b3/s2_2_3.html.

Notes
• Ranking is based on gross disbursements.
• Due to rounding, the total may not match the sum of each number.
• [–] indicates that no assistance was provided.
• Grant aid includes aid provided through multilateral institutions that can be classified by country.
• Aid for multiple countries is aid in the form of seminars or survey team dispatches, etc. that spans over multiple countries within a region.
• Country or region shows DAC recipients but including graduated countries in total.
• Negative numbers appear when the recovered amount of loans, etc. exceeds the disbursed amount.

"Partnership for Quality Infrastructure", which is analyzed in the next section, while employing the strategy of "packaged exports", which is explained in detail in the sections below.[25] Essentially, this has been an attempt to encourage Japanese corporations to invest more in Asia in general (and not only in CA), while the Japanese government provides financial and conceptual support for such engagements. In real terms, the concept of "Partnership for Quality Infrastructure" is meant to send a message to potential partners that Japanese infrastructure know-how might be more expensive when compared to other available alternatives (including those of China), but it provides quality and standards that might be worth the money spent on it. Therefore, this initiative can also be interpreted as an attempt by the Japanese PM to turn the weaknesses of Japanese infrastructure development projects, namely, their high costs, into a competitive advantage by emphasizing unrivaled Japanese quality and impeccable standards. The Japanese (unmentioned) "other" in this initiative is China, with its cheap and fast infrastructure projects that have not always been received with satisfaction because of issues and concerns with quality and standards. However, there are few examples of the application of this concept to Japanese infrastructure development abroad from which to draw any conclusions.

In terms of the security-related frames employed in narrating their projects, the definition of "security" in the cases of Chinese and Japanese infrastructure development projects have different meanings. In the Chinese government's interpretation, the notion of security has been closely related to economic cooperation, of which infrastructure development is considered to be a part.[26] Therefore, for the Chinese government, economic cooperation involving the construction of large and small infrastructure projects means providing economic opportunities to the receiving party while also contributing to the overall improvement of the security situation. In this case, the Chinese interpretation of security implies fighting insecurity, identified as terrorism, separatism and extremism. Any project that brings about development for the Chinese government has the implication of also addressing the socio-economic roots of terrorism and crime. Therefore, for many experts in China, the issues of economic infrastructure development cannot be separated from those of fighting terrorism. Logically, the infrastructure development projects that connect China (and in particular, its unstable region of Xinjiang) to CA are framed in the official Chinese government discourse as bringing about development, which in turn brings stability.

Japanese infrastructure development in CA is vaguely connected to Japan's security concerns. The Japanese government's official security discourse related to infrastructure development is generally connected to the need for assistance in fighting drug trafficking and the notions of rebuilding and pacifying Afghanistan.[27] Frequently, direct and indirect (through international organizations such as the Asian Development Bank, or ADB) support for infrastructure development – such as that exemplified by the Central Asian Regional Cooperation initiative – attempts to facilitate better conditions for connectivity between Afghanistan and its CA counterparts to promote the economic development of Afghanistan and thus provide solutions to social and economic problems in that country. The logic of Japanese support for such a project is somewhat similar to that behind Chinese support for developmental projects; they bring stability, which then brings security. However, the biggest difference between the Chinese and Japanese approaches is that Japan emphasizes the notion of human security – the notion that received support in Japan in the mid-1990s. To further this goal, Japanese assistance has been extended to other projects that provide food security and water supply security.[28] These projects have been more focused on the demographically bigger countries of Kazakhstan and Uzbekistan, while certain programs have also been maintained in the smaller countries of Kyrgyzstan and Tajikistan.[29] For the Japanese government, any infrastructure development needs to be centered on the concept of human security and the creation of secure and sustainable livelihoods for the populations living where the infrastructure-related projects are built. This notion has also been reflected in the High Quality Infrastructure Partnership initiative. In addition, such infrastructure development often entails the development of human capacity, a component that is emphasized by Japanese practitioners and policy makers as reflecting the "Japanese-ness" of Japan's developmental assistance.

Trade partner *vs.* aid provider

As demonstrated above, there are several instances in which the framings of Chinese and Japanese foreign policies – and of their infrastructure development projects in CA – differ from one another. However, the biggest difference can be seen in the different roles each country plays when implementing its infrastructure development projects. China and Japan prioritize different areas of their projects and use different strategies in approaching them. For China, several areas are defined as the highest priorities for Chinese corporate and state interests (as indicated below and in Chapter 6 on Chinese road maps for Uzbekistan). Even before the BRI initiative was announced, Chinese corporations were actively involved in the extraction of mineral resources and in the creation of energy pipelines for exports of these resources into China. Kazakhstan, Uzbekistan and Turkmenistan were prioritized in these types of projects because of the rich resources they possess and their interest in developing alternative (to Russia) markets for their resources. What followed was the construction of transport corridors, which included not only traditional energy resource export pipelines but also roadways and railroads to connect Chinese producers with markets in CA and with transit areas to bring their goods further into Russia, Europe and other parts of the world. In addition to constructing new corridors, Chinese infrastructure development also involved attempts to connect already existing corridors with the new ones.

In such a structure, energy resource–related projects are of primary importance to China because of domestic demand for energy-related products, the resource-based structures of CA economies, and the close geographic proximity of CA to Chinese consumers. These projects allow energy resources to be delivered to end-users in a relatively rapid manner and at lower transportation costs compared to other alternative energy resource producers. Moreover, projects to create trade hubs in the areas bordering Kazakhstan, Kyrgyzstan and Tajikistan were also attributed high priority for China because they were intended to connect producers in China with consumers in CA. There were also several transportation infrastructure projects, such as tunnel construction in Uzbekistan, involving the creation of transportation facilities with technology not available to CA counterparts. In such projects, the main construction force has been the Chinese construction companies and a workforce brought into CA from China. These projects signified breakthroughs in certain areas – and had profit-generating potential – for CA states because the transportation hubs are considered to intensify economic intra-state and inter-state trade within CA. Furthermore, there were a few instances in which projects related to energy and resources were implemented using the land-for-infrastructure schemes, with a few infamous ones in Tajikistan and Kyrgyzstan. These projects were specifically implemented in economically less endowed countries (Kyrgyzstan and Tajikistan), where financing for infrastructure has not been available and where certain plots of land were transferred to China using a long-term lease. However, such instances have been rare and have been met with public disapproval in the countries involved. This has also led to the abandonment of "land-for-infrastructure" practices, while the Chinese foreign ministry officially plays down the prospects for such deals in CA.

In terms of funding for these projects, several financial instruments have been used by the Chinese government to generate financial support. Among these instruments, those most committed to financing infrastructure projects are Eximbank, along with the newly created AIIB, the Silk Road Fund and the SCO bank. The financial resources provided by these institutions, combined with Chinese construction know-how and the relatively low cost of various infrastructure-related projects, serve as powerful instruments in promoting Chinese-led projects in CA.

Over the years since the CA republics' independence, Japan has allocated a massive amount of ODA financial assistance toward supporting the independence of these states.[30] Some of this assistance has been channeled to infrastructure development projects, including improvement of water-related facilities and the quality of motorways, the construction of bridges and transboundary crossing points, and the provision of necessary equipment. In particular, the construction and provision of border crossing points between Afghanistan and its CA counterparts have been Japan's most significant contributions to promoting trade and fighting drug trafficking. The provision of equipment for customs controls allows shortening the time required for checks at the borders and allows more thorough checks to be conducted.[31] Without such facilities, transportation through the borders required a considerably longer time for controls, while the efficiency of customs inspections was rather low. Other projects included the construction of solar and conventional energy generation plants.[32]

Another aspect of Japanese involvement in infrastructure development in CA relates not only to the creation of new energy-related infrastructure in the region but also to the maintenance and modernization of Soviet-era infrastructure.[33] This represents one of the most important aspects of infrastructure development in CA because, although the basic infrastructure in CA states was well constructed in the Soviet era, many elements of the infrastructure created by Soviet planners are now aging and require urgent modernization. Modernizing the existing infrastructure is just as important as creating new infrastructure.[34] Japan possesses expertise and know-how in this area, and Japanese corporations are often the winners of bidding contests for projects focusing on the modernization and maintenance of energy generation plants throughout CA as well as projects for water pumping stations.[35]

In terms of new infrastructure creation, as if to compensate for the higher costs and lack of financial commitment from Japanese financial institutions for projects in CA, the Japanese PM announced the Partnership for High Quality Infrastructure in 2015.[36] Although this initiative is not CA-focused and is not directly linked to Chinese international expansions, the Japanese PM is certainly concerned with increasing the competitiveness of Japanese infrastructure development projects internationally. To promote Japanese strengths in such projects, the Japanese government emphasizes that Japanese involvement in projects is motivated not only by the gains Japan stands to receive from them but also by certain standards, such as high quality and the long-term needs of the receiving country. Although the costs of Japanese infrastructure projects are sometimes higher than the costs of available alternatives, the Japanese PM emphasizes that such costs are attributed to the fact that Japan has high standards, such as economic efficiency (which means that the receiving party is not over-indebted as a result of implementing a project), safety (including human security and security of livelihood), resilience to natural disasters (based on the experience Japan has acquired as a natural disaster–prone country), and consideration for the environmental and social costs of such projects. This point is of special importance because Japan is at the forefront of the development of environmentally friendly technology and is also frequently cited as a better infrastructure development partner than China or other available alternatives. Finally, Japan emphasizes that its infrastructure development projects contribute to local society in terms of both access to infrastructure and the transfer of technology/know-how.

To operationalize these principles, the Partnership for High Quality Infrastructure emphasizes a four-fold mechanism that largely builds on the assistance-extending expertise and experience of the Japanese government. This includes but is not limited to the expansion of assistance through Japan International Cooperation Agency (JICA) projects, collaborating with the ADB for financial assistance, funding high-risk projects through the Japan Bank for International Cooperation (JBIC) and other risk-taking financial institutions, and, importantly, setting the standards of the Japanese Partnership for Quality Infrastructure as the international standards for infrastructure project implementation.[37]

Contested railroads? Chinese railroads and their Japanese alternative

As mentioned above, one of the most ambitious proposals of the Chinese government with respect to the CA region is the proposed construction of the "Silk Road Economic Belt" BRI concept (consisting of six economic corridors, of which the Eurasian Land Bridge, China–Central Asia–West Asia, and the "Twenty-First Century Maritime Silk Road" are relevant to CA).[38] Because this proposal required a financing arm, the AIIB was established to secure a stable and consistent source of funding for BRI.[39] While the plans for such transport corridors have been discussed at the inter-state committees between China and its CA counterparts, one of the breakthroughs was achieved during the visit of the newly elected president of Uzbekistan to China in May 2017. Among the great number of agreements signed during the visit, several (railroad infrastructure development for US$520 million and Tashkent–Osh road construction for US$220 million) were related to transportation infrastructure development. These were part of the agreement between Uzbekistan and the PRC to facilitate smooth international road transportation between the countries, which involves the simplification of procedures and the creation of an environment to increase the transportation of goods using land roads.[40]

This project stipulates transportation infrastructure (rail and motorways) construction between the Uzbek city of Andijan and the Chinese city of Kashgar, with a route going though Kyrgyz Osh and Irkeshtam. This is the shortest route from China to Uzbekistan, and both countries are interested in its construction.[41] China has called for such railroad construction for several years as a way to connect China with other markets in Europe through CA's transport networks. For Uzbekistan, this represents the shortest way to transport its goods into China by avoiding Kazakh railroads, which result in a longer transportation period and higher costs.[42] Kyrgyzstan also announced that construction of such a road will free it of dependence on Kazakh and Russian railroads for transporting its goods. This announcement was made in light of recent Kazakh–Kyrgyz tensions regarding the alleged interference of Kazakhstan in the Kyrgyz election process in 2017.

Uzbek president Islam Karimov lobbied for this project on several occasions. Governments of China and Uzbekistan discussed this project back in 1992 during the visit of Qian Qichen, foreign minister of China at that time, to Uzbekistan. In 1994, President Karimov raised the importance of construction of a direct railway from Uzbekistan to China through Kyrgyzstan during PM Li Pen's visit to Uzbekistan.[43] In 1998, China signed an agreement with Uzbekistan and Kyrgyzstan on the construction of this road and motorway.[44] It would allow China to shorten the distance to transport its goods from China by avoiding Kazakh railroads. It would also decrease the dependence of China on Kazakh railroads, thus pre-empting difficulties in transporting goods if relations between China and Kazakhstan were to worsen in the future.

Prior to the BRI announcement, in 2012, Kyrgyzstan drafted its own railroad project along this route, which was supposed to be 380 km longer than the current one. Kyrgyzstan attempted to create a railroad system that would not only connect China to Uzbekistan through the shortest route but also cover remote areas of Kyrgyzstan currently not connected to the national railroad system.[45] However, Uzbekistan and China objected to such a route change, as it would imply losses in terms of time for transporting goods and costs associated with construction.[46] With the warming of Uzbek–Kyrgyz relations in late 2016 and 2017 and the crisis in Kyrgyz–Kazakh relations in 2017, the Kyrgyz president announced that Kyrgyzstan will prioritize construction of this road.

In addition to railroad infrastructure, a motor road connecting China to Uzbekistan through Kyrgyzstan is also planned for construction. The border crossing point of Irkeshtam is located 240 km from Kyrgyz Osh and is within 285 km of Uzbek Andijan. The distance from Irkeshtam and Kashgar is 250 km. The construction of these railroad networks is prioritized by China and its CA counterparts such as Uzbekistan and Kyrgyzstan. However, there are certain challenges which include, but are not only related to, the route to be used for such construction, technicalities of the transit of goods through Kyrgyzstan and the financing of the project, especially its Kyrgyz part.

In addition to railroad construction, the Chinese Railway Tunnel Group, which built the Kamchik Tunnel in Uzbekistan, has also committed to the construction of a motorway under the Kamchik Tunnel for vehicles, which is called project Kamchik 2.[47] The railroad and motorways mentioned above are intended to increase connectivity between CA and China and create new transportation infrastructure that is currently non-existent.

Japan's stated goal for involvement in this area is somewhat similar to China's: aiming to assist CA countries in modernizing their infrastructure. In this sense, the Japan External Trade Organization (JETRO) maintains that the infrastructure construction projects supported by the Chinese government – such as the China Land Bridge (CLB) and the Trans-China Railway (TCR) – do not necessarily collide with Japanese intentions in this region but, on the contrary, may have the effect of enhancing trade between CA and Japan through seaports in China. JETRO, in particular, also emphasizes that existing railroads do not necessarily serve the interests of developing relations between CA and Japan, and thus additional infrastructure, even that financed by China, can support connectivity between Japan and CA.[48] However, as described below, Japanese participation in the railroad infrastructure projects aims to modernize existing infrastructure as opposed to creating new infrastructure from scratch. Such an approach relates to many factors described in the sections above, such as the ambiguity of CA's importance for Japan, its relative distance from the CA region, limited Japanese corporate penetration of CA, and a lack of massive financial resources comparable in scale to the Chinese resources to be spent in the CA region.

Until the death of the first Uzbek president Karimov, the government of Uzbekistan emphasized infrastructure projects that supported the independence of the Uzbek railroad system from other countries. This was also the tendency of the support provided by JICA to, for instance, the Karshi–Termez railroad electrification project in Uzbekistan so that Uzbek goods would not have to cross the borders of Turkmenistan while being transported within Uzbekistan's own territory. Moreover, the support provided by JICA aimed to modernize infrastructure that was constructed in this region during the Soviet era. CAREC (Central Asia Regional Economic Cooperation) and TRACECA (Transport Corridor Europe–Caucasus–Asia) are excellent examples of Japanese support for modernizing and constructing transport-related infrastructure in this region.

The major problem with the Chinese railroad projects – a problem emphasized by both Japanese and CA experts – is that the infrastructure projects did not lead to the accumulation of infrastructure construction know-how in the countries where the projects were implemented. Therefore, the Japanese have attempted to learn from Chinese involvement and offer not only technology but also training so that the projects can later be maintained by CA specialists.

Japanese infrastructure involvement in this region does take the form of direct involvement by the Japanese government through its ODA program, with JICA as its main actor; there is also support for the CAREC and TRACECA projects. A recent attempt by Japanese companies to expand the scale of their overseas infrastructure involvement occurred in 2010 under the "Package-gata kaigai tenkai" [packaged infrastructure overseas expansion] scheme, supported by the office of the PM to stimulate the expansion of Japanese corporations into Asia-centered international infrastructure projects.[49] One of the biggest aims of this scheme was to promote exports of technology and hardware related to the construction of railroad infrastructure, as Japan is rightly famous for the safety and reliability of these products. Such expansion aims at the exports of not only high-speed train technology but also conventional railroad infrastructure construction, for which cooperation has been established among the Ministry of Foreign Affairs, Ministry of Economy and Industry and Ministry of Land, with JBIC as the main financial institution.[50] Moreover, a company called Japan Consultants was established in 2011.[51] The projects involve not only the construction and provision of technology but also – importantly – maintenance after the project has been completed.

In the majority of cases, infrastructure expansion involves packaged exports (installing the hardware and software), exports of integrated systems (exports of not only the infrastructure of railroads but also complementary infrastructure such as railroad lights and other supplementary equipment), and exports with operational obligations (implies not only installing the previous two elements but also committing to operating the system after it is in place). Japan now aims to actively use the first type of involvement (referred to as packaged infrastructure exports) through JICA and to use its capacity to recognize the needs of other developing countries and offer them such exports in place of ODA assistance.[52]

Infrastructure-related projects in CA supported by the Japanese ODA can be effectively represented by Japanese involvement in Uzbekistan. In Uzbekistan, the particular railroad infrastructure project supported by Japan is the Tashguzar–Kumkurgan railroad modernization.[53] The goal was to enhance the railroad's capacity to facilitate the transit of a larger number of trains; the project was launched in 2004 and completed in 2010. Although Japan did construct a completely new rail line linking regions of Kashkadarya and Surkhandarya of Uzbekistan, this project was aiming at rehabilitating the existing railway line between Karshi and Tashguzar in the Kashkadarya region and promoting transport goods without having to cross the border of Turkmenistan.[54] Thus, although Japan does construct smaller sections of railroads, these are connected to projects for the rehabilitation of existent infrastructure and are generally not as large as Chinese infrastructure construction.

In addition, the Karshi–Termez railroad has been modernized (by the electrification of some railroad parts), followed by the modernization of the Tashkent–Fergana railroad, which required technology that Uzbekistan did not possess at that time and which was provided by Japanese corporations.[55] Japan has also provided Uzbekistan with the software that enabled the latter to develop a more effective schedule for railroad functioning, which is considered to be a type of know-how previously unavailable in Uzbekistan. Nevertheless, the problem with this provision of technology to CA states, and Uzbekistan in particular, is that these countries receive a large amount of support for infrastructure development from Japan, China and Germany. However, the degree of complementarity among the different technologies is not high, thus creating problems with maintenance. In addition, the software used by the majority of CA states in creating their railroad schedules is mainly based on old Russian technology such as "RIXT", which is not modernized and does not have a high degree of complementarity with the technology being provided by other countries.[56]

As has been seen, China and Japan are actively engaging in infrastructure development in CA. However, they see their roles and contributions differently, with China attempting to connect CA infrastructure to its own and Japan aiming to export its technologies and providing expertise in modernizing the available infrastructure mainly through the ODA assistance scheme.

The case of Japanese energy-related infrastructure development in Uzbekistan

In addition to transportation infrastructure development, another area of active Chinese and Japanese engagement is energy-related infrastructure development. The issues and patterns of cooperation between CA countries and China and Japan are also well exemplified by the case of their interactions with Uzbekistan. Chinese interest in this energy-rich CA country relates to three types of engagements: energy-based resource export infrastructure (from Uzbekistan to China), new extraction infrastructure and the creation of an energy resource processing sector of the economy.[57]

The most recent and the largest agreements between Uzbekistan and China have been those focusing on the joint production of synthetic fuel (US$3.7 billion), investing in Uzbekistan's oil industry (US$2.6 billion), and agreements on cooperation in the construction of energy generation plants (US$679 million).

Among Uzbekistan's exports to China, mineral and natural resources constitute a considerable share of the trade between the countries. According to agreements concluded in May 2017 during Mirziyoyev's visit to China, contracts identified natural gas (six billion cubic meters worth US$734 million), uranium (US$30 million), textiles (US$200 million), leather (US$21.3 million), and agricultural products (US$1.6 million) as products to be exported to China by the end of 2017. In addition, plans have been articulated for additional exports of natural gas to reach US$2.4 billion for the years 2018 to 2020.[58]

There are a few plans for the construction of additional natural gas pipelines to connect Turkmenistan, Uzbekistan and Kazakhstan to Chinese consumers. However, these discussions have not yet materialized into specific construction projects or financial commitments because of questions regarding the economic sustainability of the pipelines' operations.

In terms of new extraction rights, CNPC (the China National Petroleum Corporation) secured the co-financing contract with the Bank of China for a drilling project at the gas condensate field in Bukhara by establishing JV New Silk Road Oil and Gas, which was set up by Uzbekneftegaz (UNG) and China's CNODC (China National Oil and Gas Exploration and Development Corporation, a subsidiary of CNPC).[59] According to the license granted to the joint venture, it plans to develop the existing wells and drill another 16, with annual production to reach one billion cubic meters of natural gas and 6,500 tons of condensate.[60]

In terms of the generation of new industries, Uzbekistan concluded an agreement between UNG and the Chinese Development Bank (worth US$3.7 billion, of which US$1.2 billion is to be financed by China) to finance the establishment of a plant to produce synthetic fuel at Uzbekistan's largest gas refinery complex, Shurtan.[61] Accordingly, the plant is intended to process 3.6 billion cubic meters of natural gas into 743,500 tons of synthetic fuel, 311,000 tons of aviation fuel, 431,100 tons of naphtha fuel and 20,900 tons of liquefied gas.[62] Interestingly, technological support for the plant is to be provided by South Korea's Hyundai Engineering & Construction under a license provided by South Sasol. The technology for turning natural gas into liquefied gas is provided by the Dutch firm Haldor Topsoe.

Hydro-energy generation has also been an area of Chinese interest, with a US$3 billion agreement between the Ministry of Commerce and Uzbekgidro signed in May 2017 to install and use approximately 300 water pump stations for electricity generation.[63] This is also in line with Uzbekistan's own strategy of hydro-energy development adopted in November 2015, which aims to invest US$889.4 million into this sector between 2016 and 2020.

The modernization of the current energy-generating capacity of Uzbekistan has also been prioritized in negotiations. In particular, China Railway Tunnel Group (CRTG) and China Coal Technology & Engineering Group began the modernization of a coal extraction plant to achieve extraction levels of 900,000 tons of coal per year, with the amount of investment equaling US$94.5 million.[64] In addition to the modernization of plants generating traditional sources of energy, non-traditional sustainable sources such as biomass generation have also been the subject of agreements. Uzbekneftegaz, AKB Agrobank and China's

Poly International Holding signed a memorandum of cooperation to establish the production of modern biogas plants worth US$10 million and to assist in the modernization of eight domestic enterprises, including the JSC Oil and Gas and Chemical Engineering Plant, in line with the governmental Program of Measures to Increase Biogas Plants in Uzbekistan for 2017–2019.[65] As seen from the areas and projects described above, China is committed to investing in economically sound project infrastructure and sectors of the economy that did not previously exist in the country. This reflects the economic might of China and the expanding nature of its corporate interests.

Japanese companies and state agencies do not commit to engagements in the way that China does, as already explained. As mentioned by scholars, "ODA projects are not foot-in-the-door pathways for Japanese involvement in Central Asia" but rather "are the bulk of their business operations and sometimes the only raison d'être for regional presence".[66] Thus, Japan (the government, its affiliated agencies and corporations) is keener on using Japanese expertise that is not available in China or elsewhere to both cement its presence in this region and emphasize the "Japanese-ness" of the assistance schemes. At the same time, the major difference from the Chinese approach is that Japanese involvement in this region is largely led by the initiatives of the Japanese government and not the interest of private enterprises, which lack information on CA markets and the confidence to act in "new waters".

Among the many issues, the Japanese developmental agencies, most prominently JICA, tend to emphasize the aging of the energy supply infrastructure and its inefficient functioning as the problem leading to losses of already scarce energy resources before they reach end-users. Thus, Japan, in its CA engagements in general and in Uzbekistan in particular, contributes to the modernization of such infrastructure because it possesses the needed technology and because the task does not require the scale of investments required to launch new sectors of the economy, as exemplified by China's involvement.

The case of Japanese involvement in the energy sector of Uzbekistan is again symbolic in this respect. For instance, in a drive to liberalize the market for energy resources, the government of Uzbekistan shifted the power over energy supply from the Ministry of Energy to the state company Uzbekenergo in August of 2001.[67] Accordingly, the JICA reported that according to its estimates, 85 percent of the energy provided by 46 thermal plants in Uzbekistan was generated from thermal power, while only 12 percent of the energy was from water-related resources.[68] However, the majority of thermal plants constructed and functioning in the country were built during the 1960s and 1970s and had been aging; they were considered to be functioning at 60 percent of their capacity. In particular, the turbines of the thermal plants were subject to severe effects of aging, and urgent action was required to modernize them. The second problem was related to the fact that this aging energy supply infrastructure was not only inefficient but also produced a massive amount of CO_2 gas emissions in relation to GDP, placing Uzbekistan among the world's leading polluters.[69] The problem of energy supplies has been the most acute in the Fergana Valley, where the population density is highest. Modernization and construction of thermal energy generation infrastructure in this part of the country were calculated to provide an 8 percent increase in the energy supply to end-users.[70]

In terms of the JICA support for the modernization of thermal power plants, the announced goal was to support energy infrastructure modernization in terms of both hardware (meaning provision of technology and equipment) and software (meaning provision of training and development of human capabilities to maintain the system).[71] While JICA acknowledges that it lacks the resources to provide for the projects on a scale equal to that of China, JICA emphasizes the specialized nature of its contribution to energy-related infrastructure development by providing two-fold assistance: modernizing existing infrastructure and providing training to enable local technicians to run the system sustainably in the long run. In terms of hardware support, JICA has assisted by providing turbines in 2010 (under a co-funding scheme with the ADB) and 2013 (under the ODA scheme) to modernize the Talimarjan and Navoi thermal power plants, respectively.[72] Such assistance provides not only a renovation of existing facilities but also a more efficient technology, which enables the use of thermal energy with a higher degree of efficiency.[73] Similarly, the modernization of the Turakurgan facility – capable of 800 MW – was launched in November of 2014.[74]

Turbines provided as part of the ODA have been produced by Mitsubishi Hitachi Power Systems (MHPS) and used at the Navoi 1 plant. In October 2016, MHPS and Mitsubishi Ltd provided the turbine for the Navoi 2 plant.[75]

In providing such equipment, Japanese companies use the following criteria to evaluate the environments in which they operate. According to an interview with representatives of Tashkent, for MHPS, there are three main criteria used to evaluate the operational environment: first, the existence of a need (large population and aging infrastructure); second, the existence of a pro-Japanese environment in the country, and third, a high degree of literacy in the region.

In terms of the challenges encountered by the Japanese corporations, one aspect is the difficulty of delivering Japanese technology to end-users because the guidelines and manuals are written in Japanese. Although the Japanese government has invested large amounts of resources into encouraging Japanese language education, the level of Japanese literacy required to read technical documentation has been too high, thus requiring many interpreters.

In terms of approach, the Uzbek government supports the Japanese approach because it generates both technology transfer and employment. It also contrasts with the Chinese approach, which favors bringing hardware and a workforce into the country, as exemplified by Chinese engagement in Kyrgyzstan. Uzbekistan did not allow this approach to be applied and thus continues to favor the Japanese assistance scheme.[76]

Conclusions

As outlined above, both China and Japan consider CA to be one of the new frontiers for developing their infrastructure and expanding their corporate interests.

Although their areas of interests largely overlap and are mainly focused on energy and transportation infrastructure, there are significant differences in how these countries justify, frame and narrate their involvement in infrastructure development in this region. The countries use similar strategies of contributing to CA regional development by providing expertise and infrastructure to assist these states' decolonizing agenda. Both China and Japan attempt to define the importance of CA regional engagement not only through the lens of external expansion but also domestically. For China, such importance is related to connecting its producers to international markets, while for Japan it is about the expertise it possesses domestically, which can contribute to international development and impact domestic growth. Thus, in the Chinese case, such redefining of CA's importance is obviously connected to constraints of domestic development of Xinjiang and other areas; in the case of Japan, CA serves as a good example to demonstrate how Japan adapts domestically (through ODA, Quality Infrastructure Initiative etc.) to adjust to the new role it needs to play in international affairs. Thus CA plays the role of a new frontier for such domestic adaptation in the Japanese case.

There are striking differences in the approaches and roles of each of these states. Close geographic proximity is a significant advantage for China in arguing for the intensified construction of infrastructure that connects Chinese producers and CA consumers. In addition, energy and transport infrastructure offers CA states an alternative for exporting their raw mineral resources. As seen from the list of projects above, Chinese-led energy and transport infrastructure projects are gradually showing signs of spill-over into other areas, and some projects now include the construction of assembly lines in CA. In addition, the complementarity of Chinese projects with other states (Russian-led Eurasian Economic Union and others) serves China well for now because the Chinese government backs up its corporate interests through offers to insure the risks taken by Chinese corporations and by providing financing for infrastructure development. Moreover, with an increase in the Chinese presence and the number of infrastructure-related projects, there is also an increasing concern among CA states regarding China's utilitarian approach to CA, as is demonstrated by the case of Kyrgyzstan. This could be among the grand challenges to further Chinese penetration in this region.

The Japanese approach to CA in general and to infrastructure development in particular differs significantly from China's. Japan attempts to build its competitive advantage by emphasizing the notions of (a) human security-centered infrastructure and (b) more advanced technology. In addition, Japanese support for smaller infrastructure development projects (such as agriculture, community building, and customs) outside the energy-related fields received a warm welcome in most CA countries. Further, Japan's lack of geographic proximity can also serve as a positive feature and even as a competitive advantage for the broader participation of Japanese corporations in infrastructure development because this contrasts with China's assumed neo-colonization. However, there is a need for Japan to move away from the emphasis on energy-related infrastructure/imports of energy resources because the geographic distance between Japan and CA complicates the task of transporting CA energy resources. Finally, the Japanese promotion of quality infrastructure represents an adjustment to both the competitive environment for new opportunities and Japan's attempt to frame its more expensive technology as competitive advantage.

Notes

1 Timur Dadabaev, "Chinese and Japanese Foreign Policies towards Central Asia from a Comparative Perspective", *The Pacific Review* 27, no. 1 (2014): 123–145, doi:10.10 80/09512748.2013.870223; Timur Dadabaev, "The Chinese Economic Pivot in Central Asia and Its Implications for the Post-Karimov Re-emergence of Uzbekistan", *Asian Survey* 58, no. 4 (2018): 747–769, doi:10.1525/as.2018.58.4.747; Timur Dadabaev, "'Silk Road' as Foreign Policy Discourse: The Construction of Chinese, Japanese and Korean Engagement Strategies in Central Asia", *Journal of Eurasian Studies* 9, no. 1 (2018): 30–41, doi:10.1016/j.euras.2017.12.003; Timur Dadabaev, "Engagement and Contestation: The Entangled Imagery of the Silk Road", *Cambridge Journal of Eurasian Studies* 2018, no. 2, doi:10.22261/cjes.q4giv6.

2 M Auezov, "Kazakhstan Must Stop Wavering between Russia and China to Pursue Central Asian Consolidation", *Interfax*, January 29, 2013; M Auezov, "Ex-Ambassador of Kazakhstan to China Concerned over China's Classified Documents", *Tengri News*, 2015, www.en.tengrinews.kz/politics_sub/Ex-Ambassador-of-Kazakhstan-to-China-concerned-over-Chinas-21645/, last accessed on January 15, 2016.

3 "V besporyadkah v Toguz-Torunskom Raione zameshany politiki, imena kotoryh uzhe izvestny: PM Iskakov" [PM Iskakov: Certain politicians whose names are already known are part of the riots in Toguz-Torunsk region], *Kyrtag (Kyrgyz Telegraph Agency)*, April 13, 2018, https://kyrtag.kg/ru/news-of-the-day/v-besporyadkakh-v-toguz-torouskom-rayone-zameshany-politiki-imena-kotorykh-uzhe-izvestny-premer-isak.

4 Timur Dadabaev, "Shanghai Cooperation Organization (SCO) Regional Identity Formation from the Perspective of the Central Asia States", *Journal of Contemporary China* 23, no. 85 (2014): 102–118, doi:10.1080/10670564.2013.809982.

5 Tony Tai-Ting Liu, "Undercurrents in the Silk Road: An Analysis of Sino-Japanese Strategic Competition in Central Asia", *Japanese Studies* 8 (March 2016): 157–171.

6 N Swanström, "China and Central Asia: A New Great Game or Traditional Vassal Relations?", *Journal of Contemporary China* 14, no. 45 (2005): 569–584, doi:10.1080/10670560500205001; T Uyama, "Shin 'Great Game' no Jidai no Chuou Ajia" [Central Asia in the era of new "Great Game"], *Gaiko* 34 (2015): 18–20; A Cooley, *Great Games, Local Rules: The New Great Power Contest in Central Asia* (Oxford: Oxford University Press, 2012).

7 Dadabaev, "Chinese and Japanese Foreign Policies"; Dadabaev, "'Silk Road' as Foreign Policy Discourse".

8 S Ramani, "China's Expanding Security Cooperation with Tajikistan", *The Diplomat*, July 16, 2016, http://thediplomat.com/2016/07/chinas-expanding-security-cooperation-with-tajikistan/.

9 S Zhao, "Foreign Policy Implications of Chinese Nationalism Revisited: The Strident Turn", *Journal of Contemporary China* 22, no. 82 (2013): 535–553.

10 Shanghai spirit implies norms that connect the issues for cooperation prioritized by both countries without seeking unilateral gains. These norms also imply the importance of mutual sacrifices and compromises for mutual gains.

11 Office of the Leading Group for the Belt and Road Initiative, *Building the Belt and Road: Concept, Practice and China's Contribution* (Beijing: Foreign Languages

Press, May 2017), pp. 11–17, https://eng.yidaiyilu.gov.cn/wcm.files/upload/CMSydylyw/201705/201705110537027.pdf, accessed March 12, 2018; National Development and Reform Commission, Ministry of Foreign Affairs and Ministry of Commerce of People's Republic of China, "Vision and Actions on Jointly Building Silk Road Economic Belt and 21st-Century Maritime Silk Road", March 30, 2015.

12 H Yu, "Motivation behind China's 'One Belt, One Road' Initiatives and Establishment of the Asian Infrastructure Investment Bank", *Journal of Contemporary China* 26 (2017): 353–368, doi:10.1080/10670564.2016.1245894.

13 Y Wang, "Offensive for Defensive: The Belt and Road Initiative and China's New Grand Strategy", *The Pacific Review* 29, no. 3 (2016): 455–463, doi:10.1080/0951274 8.2016.1154690.

14 As an example of such new engagement, see Ministry of Foreign Affairs of the People's Republic of China, "Joint Declaration on New Stage of Comprehensive Strategic Partnership Between the People's Republic of China and Republic of Kazakhstan", August 31, 2015, www.fmprc.gov.cn/mfa_eng/wjdt_665385/2649_665393/t1293114.shtml, last seen on January 25, 2018).

15 R Hashimoto, Address to the Japan Association of Corporate Executives, Tokyo, July 24, 1997, www.Japan.kantei.go.jp/0731douyuukai.html.

16 *ODA Charter* (Tokyo: Ministry of Foreign Affairs of Japan, 2003), www.mofa.go.jp/policy/oda/reform/charter.html.AII.

17 Japan International Cooperation Agency (JICA), *JICA fusei fuhai boushi guidance* [Guidance on prevention of corruption] (JICA, October 2014), www2.jica.go.jp/ja/odainfo/pdf/guidance.pdf.

18 L Sun. "Uzbekistan I Kitaj gotovy prodvigat stroitel'stvo novogo Evrazijskogo kontinental'nogo mosta" [Uzbekistan and China are ready to promote construction of new Eurasian land bridge], Interview with Ambassador of China to Uzbekistan Sun Lijie, *Podrobno*, June 2, 2017, www.podrobno.uz, last seen on June 8, 2017.

19 D Mitchell and C McGiffert, "Expanding the 'Strategic Periphery': A History of China's Interaction with the Developing World", in *China and the Developing World: Beijing's Strategy for the Twenty-First Century*, ed. J Eisenman, E Heginbotham and D Mitchell (New York: ME Sharpe, 2007), pp. 3–25, especially pp. 7–9.

20 On the evolution and deconstruction of "win–win" principle, see Rosita Dellios, "Silk Roads of the Twenty-First Century: The Cultural Dimension", *Asia & the Pacific Policy Studies* 4, no. 2 (2017): 225–236, doi:10.1002/app. 5.172.

21 For examples, see Dadabaev, "The Chinese Economic Pivot".

22 For the latest protests and claims, see "V besporyadkah v Toguz-Torunskom Raione zameshany politiki", *Kyrtag*. For previous debates on this see Timur Dadabaev, "Japan's Search for Its Central Asian Policy: between Idealism and Pragmatism", *Asian Survey* 53 (2013): 506–532, doi:10.1525/as.2013.53.3.506.

23 N Murashkin, "Japanese Involvement in Central Asia: An Early Inter-Asian Post-Neoliberal Case?" *Asian Journal of Social Science* 43 (2015): 50–79, especially p. 56.

24 T Uyama, "Japanese Policy in Relation to Kazakhstan: Is There a 'Strategy'?", in *Thinking Strategically: The Major Powers, Kazakhstan, and the Central Asian Nexus*, ed. R Legvold (Cambridge, MA: MIT Press, 2003), 165–186, especially p. 178.

25 S Abe, "The Future of Asia: Be Innovative" (speech at the 21st International Conference on the Future of Asia, May 21, 2015), http://japan.kantei.go.jp/97_abe/statement/201505/0521foaspeech.html.

26 Ramani, "China's Expanding Security Cooperation with Tajikistan".

27 Embassy of Japan in Uzbekistan, "Assistance on Fighting Drug Trafficking from Afghanistan and Establishment of the Team for Implementation of the Plan on Fighting the Spread of Drugs from Afghanistan", 2016, www.uz.emb-japan.go.jp/itpr_ja/00_000043.html.

28 See for instance JICA's *Report on Assistance to the Republic of Uzbekistan on Increase of Agricultural Production* (JICA, 2003), http://open_jicareport.jica.go.jp/

pdf/11721420.pdf. Also see JICA, *Training Assistance to Uzbekistan and Tajikistan in Agricultural and Rural Development Report* (JICA, 2003), http://open_jicareport. jica.go.jp/pdf/11747110.pdf.

29 For the efficiency of assistance to the agricultural sector of Kazakhstan see the report by the Ministry of Agriculture, Forestry and Fisheries of Japan, *Current State and Prospects of Agricultural Sector in Kazakhstan*, Kaigai Shyokuryo repo-to, 2013, www.maff.go.jp/j/zyukyu/jki/j_rep/monthly/201310/pdf/21_monthly_topics_1310a.pdf. Also see K Nobe, *Agricultural Reforms in Uzbekistan*. Senshyu Keizaigaku Ronshyu, Senshyu University (2010).

30 JICA, "Official Development Assistance Agreement to Uzbekistan Is Signed", February 28, 2012, www.jica.go.jp/press/2011/20120228_01.html.

31 For instance, see JICA, *Preparatory Study into Riparian CA Syrdarya River Basin Management System* (Tokyo, Japan: JICA, 2010).

32 JICA, *JICA Final Report for Collection of Data for Electrification of Uzbek Railroads* (Tokyo, Japan: JICA, 2013), http://open_jicareport.jica.go.jp/pdf/12121992.pdf; JICA, *JICA Report on Extending Assistance for Turakurgan Thermal Power Station Construction Project* (Tokyo, Japan: JICA, 2013), www.mofa.go.jp/mofaj/gaiko/oda/about/kaikaku/tekisei_k/pdfs/11_anken_n05.pdf.

33 Mitsubishi Power Systems, "Memorandum Regarding Power Generation Operation System with UZBEKENERGO", 2015, www.mhps.com/news/20150202.html.

34 For emphasis of this side of assistance see JICA, *Supporting Uzbek Electricity Reforms with Both Hardware and Software* (Tokyo, Japan: JICA, 2014), www.jica. go.jp/topics/news/2013/20140313_01.html.

35 For details, see JICA, *Signing of the Official Development Assistance to Uzbekistan – EPC Contract for Combined Cycle Power Plant and Modification of Substation and Transmission* (Tokyo, Japan: JICA, 2013), www.jica.go.jp/press/2013/20130822_03.html.

36 S Abe, "The Future of Asia: Be Innovative".

37 Office of PM, "High Quality Infrastructure", May 21, 2015, www.mofa.go.jp/mofaj/gaiko/oda/files/000081296.pdf.

38 For details see Office of the Leading Group for the BRI, *Building the Belt and Road*, pp. 11–17.

39 See National Development and Reform Commission, "Vision and Actions".

40 Decree of the President of Uzbekistan on Reconfirming "Agreement between Republic of Uzbekistan and People's Republic of China on Facilitating Smooth International Road Transportation", PP-3143, July 20, 2017, in Russian at http://nrm.uz/contentf?doc=508717_postanovlenie_prezidenta_respubliki_uzbekistan_ot_20_07_2017_g_n_pp-3143_ob_utverjdenii_mejdunarodnogo_dogovora&products=1_vse_zakonodatelstvo_uzbekistana, accessed September 23, 2017.

41 A Titova, "Uzbekistan hochet postroit Andijan-Osh-Irkishtam-Kashgar" [Uzbekistan wants to build Andijan-Osh-Irkishtam-Kashgar], *Kloop*, September 9, 2017, https://kloop.kg/blog/2017/09/09/uzbekistan-hochet-postroit-trassu-andizhan-osh-irkeshtam-kashgar/, accessed 23 September, 2017.

42 For details, see Dadabaev, "'Silk Road' as Foreign Policy Discourse".

43 Ablat Khodzhaev, *Kitajskij Faktor v Tsentral'noi Azii* [Chinese factor in Central Asia] (Tashkent: Fan, 2007), p. 103.

44 Khodzhaev, *Kitajskij Faktor v Tsentral'noi Azii*, p. 103.

45 M Ramtanu, "The Multiple Dimensions of China's 'One Belt One Road' in Uzbekistan", *EIR*, December 30, 2016, pp. 17–22, especially p. 20, www.larouchepub.com/eiw/public/2016/eirv43n53-20161230/17-22_4353.pdf.

46 See, for instance, Bruce Pannier, "No Stops in Kyrgyzstan for China–Uzbekistan Railway Line", *Radio Free Europe/Radio Liberty*, September 3, 2017, www.rferl.org/a/qishloq-ovozi-kyrgyzstan-uzbekistan-china-railway/28713485.html, accessed September 23, 2017.

47 "China to Help Build Second Tunnel at Kamchik Pass", *Tashkent Times*, May 16, 2017, www.tashkenttimes.uz/economy/930-china-to-help-build-second-tunnel-at-kamchik-pass, accessed September 23, 2017.

48 Japan External Trade Organization (JETRO), *Uzbekistan no Butsuryu Jijyou* [Logistical Situation of Uzbekistan] (JETRO, 2013), www.jetro.go.jp/ext_images/jfile/report/07001204/uz_logistics.pdf.

49 "Japanese Railroads Fighting Their Way in the World", *Toyo Keizai* 109 (2015): 101–200.

50 "Package-gata kaigai tenkai" [Packaged infrastructure overseas expansion], www.kantei.go.jp/jp/singi/package/pdf/hakyuukouka.pdf.

51 Japan Consultants LTD (2011), www.jictransport.co.jp/jp/aboutus/fundingcompanies/.

52 The Japanese railroad lights system prides itself on its cost-efficiency in usage because, unlike other alternatives, the Japanese signal light system is operated wirelessly and does not require connecting railroad lights by cable with one another. Also, it can not only operate with lower required maintenance costs but also can reboot and automatically restart operations more rapidly in case of emergencies. For details, see "Japanese Railroads Fighting Their Way in the World".

53 JICA, "Ex-Post Evaluation of Japanese ODA Loan Project 'Tashguzar-Kumkurgan New Railway Construction Project'", 2013, www2.jica.go.jp/en/evaluation/pdf/2013_UZB-P8_4.pdf.

54 JICA, *JICA Report on "Improving the Capacity of Functioning of the Railroads in Mountainous Areas of Uzbekistan"* (Tokyo, Japan: JICA, 2013), http://open_jicareport.jica.go.jp/pdf/1000012463.pdf.

55 JICA, *"Improving the Capacity of Functioning of the Railroads".*

56 M Koizumi, "Chiiki taikoku toshite no Uzbekistan gaikou" [Foreign Policy of Uzbekistan as a Regional Superpower] (PhD thesis, University of Tsukuba, 2018).

57 Ramtanu, "The Multiple Dimensions of China's 'One Belt One Road'".

58 "Uzbekistan planiruet k 2021 godu narastit' export gaza v Kitaj do 10 mlrd" [Uzbekistan plans to increase exports of gas to China to 10 billion cubic meters until 2021], *Interfax News Agency*, June 15, 2017, http://interfax.az/view/705802, accessed 23 September, 2017; "Uzbekistan nameren narastit postavki gaza v Kitai" [Uzbekistan aims to increase the volume of its gas exports to China], *Aziay Plus*, June 15, 2017, https://news.tj/ru/news/tajikistan/economic/20170615/uzbekistan-nameren-narastit-postavki-gaza-v-kitai, accessed September 23, 2017.

59 See "Uzbekistan–China JV New Silk Road Oil and Gas Commences Drilling in Bukhara", *Tashkent Times*, June 15, 2017, www.tashkenttimes.uz/economy/1016-uzbekistan-china-jv-new-silk-road-oil-and-gas-commences-drilling-in-bukhara, accessed September 23, 2017.

60 "Uzbekistan–China JV New Silk Road Oil and Gas".

61 George Voloshin, "Central Asia Ready to Follow China's Lead Despite Russian Ties", *Eurasia Daily Monitor* 14, no. 71 (2017), https://jamestown.org/program/central-asia-ready-follow-chinas-lead-despite-russian-ties/, accessed 23 September 2017; "Uzbekistan: President's China Trip Yields Giant Rewards", *Eurasianet*, May 18, 2017, https://eurasianet.org/s/uzbekistan-presidents-china-trip-yields-giant-rewards, accessed June 9, 2018.

62 "Uzbekistan I Kitai podpisali soglazhenij na summu bolee 20 mlrd" [Uzbekistan and China signed agreements for more than 20 billion], *Review.uz*, May 14, 2017, www.review.uz/novosti-main/item/11217-uzbekistan-i-kitaj-podpisali-soglashenij-na-summu-bolee-20-mlrd, accessed 23 September 2017.

63 "Uzbekistan I Kitai podpisali soglazhenij na summu bolee 20 mlrd".

64 "Startovala modernizatsiya predpriyatiya 'Shargunkomir'" [The modernization of "Shargunkomir" has started], *Gazeta*, September 29, 2017, www.gazeta.uz/ru/2017/09/07/coal/, accessed September 23, 2017.

65 "Program for Increased Use of Biogas in Farms Adopted in Uzbekistan", *Tashkent Times*, June 19, 2017, www.tashkenttimes.uz/economy/1060-program-for-increased-use-of-biogas-in-farms-adopted-in-uzbekistan, accessed September 23, 2017.
66 Murashkin, "Japanese Involvement in Central Asia", pp. 60–61.
67 JICA, *Republic of Uzbekistan Preparatory Survey on Turakurgan Thermal Power Station Construction Project – Final Report P15* (JICA, 2014), http://open_jicareport.jica.go.jp/pdf/12175436.pdf.
68 JICA, *Republic of Uzbekistan Preparatory Survey on Turakurgan.*
69 JICA, *Supporting Uzbek Electricity Reforms.*
70 JICA, *Republic of Uzbekistan Preparatory Survey on Turakurgan.*
71 "Turakurgan Thermal Power Station Construction Project", 2014, www.mofa.go.jp/mofaj/gaiko/oda/about/kaikaku/tekisei_k/pdfs/11_anken_n05.pdf.
72 JICA, *Supporting Uzbek Electricity Reforms.*
73 Tokyo to Kankyoukyoku [Environment Department of Tokyo], *Tennen Gasu Hatsudensho Secchi Gijyutsu Kentou Chyousa Houkokusho* [Report on Introduction of Natural Gas based energy generation system], 2012, www.kankyo.metro.tokyo.jp/policy_others/energy/docs/02%20%E7%AC%AC%EF%BC%92%E7%AB%A0GTC C%E7%99%BA%E9%9B%BB%E3%81%AE%E6%A6%82%E8%A6%81.pdf.
74 JICA, *Republic of Uzbekistan Preparatory Survey on Turakurgan*; Ministry of Foreign Affairs (MOFA) of Japan, *Seifu Kaihatsu Enjyo (ODA) Kunibetsu de-tabuku 2014 (Chuou ajia/kokasasu chiiki)* [Official Development Assistance By-country Data-book 2014] (Tokyo: MOFA, 2014), www.mofa.go.jp/mofaj/gaiko/oda/files/000072593.pdf.
75 Mitsubishi Hitachi Power Systems, "Project on GTCC Plant Construction in Navoi 2", October 19, 2016, www.mhps.com/news/20161019.html.
76 Interview with the Ambassador of Japan to Uzbekistan, March 2016.

5 Chinese, Japanese and South Korean economic cooperation road maps for Uzbekistan

Many studies have been published in recent years focusing on the foreign policy of various powers in Asia. These studies tend to focus primarily on the countries and areas in Asia that have historically received extensive attention, particularly China, Japan and South Korea in East Asia. However, few studies go beyond traditionally covered areas to focus on parts of Asia that, while becoming central to various international engagements, remain overlooked. One such example of an area not paid due attention in the literature on comparative aspects of the foreign engagements of Japan, China and Korea is what can be referred is the last "new frontier" in Asia – Central Asia (CA).[1]

The post-Soviet CA region – consisting of the five stans of Uzbekistan, Kazakhstan, Kyrgyzstan, Tajikistan and Turkmenistan – has remained marginal for Asian scholars for a number of reasons. First, this region has often been associated with the geopolitically determined larger Eurasian region consisting of Russia and other post-Soviet constituents. Thus, for many scholars in international relations (IR), this region has been approached through the analysis of Russian and post-Soviet policies, while the Asian angle of CA states' interactions has been overshadowed and to some extent hijacked by Russia-related scholarship.[2] Second, those few studies that paid attention to the CA states' interactions with Asian powerhouses, in comparative perspective, tended to focus on these states' participations in the Shanghai Cooperation Organization (SCO) or their foreign policies related to the recently announced Belt and Road Initiative (BRI).[3] Thus, once again, the framing of the CA region's coverage within the Asian political space has been hijacked by the attention paid to China-related initiatives, often with the rise of China and its global and regional economic influence as underpinning.[4] Third, those studies that intended to cover CA states' engagements with Asian countries frequently focused on individual case studies of CA–China, CA–Japan or engagements between this region and South Korea.[5] Very few studies, if any, have attempted to consider the mutual importance of CA states and powerful Asian countries in such interactions.[6] In addition, differences and similarities in Chinese, Japanese and Korean interactions have rarely been compared.[7] However, any conclusions on the role and significance of the Asian vector in foreign policies for CA states are difficult to make without an empirically grounded comparison of CA interactions with the most important and active states in this region: China, Japan and South Korea.

China, Japan and South Korea have been actively involved in various developments in CA since CA states' independence from the Soviet Union in 1991. In 1996, China launched its confidence-building mechanism, which became the Shanghai Five (three CA states, Russia, and China) and later, in 2000, the SCO.[8] These efforts resulted in significant progress in confidence-building, the resolution of border-related issues between these states, and the construction of a mechanism for combating terrorism, extremism and separatism in this region. Almost a decade later, China launched a new "Silk Road offensive" aimed at enhancing economic cooperation among the countries of the ancient Silk Road, which was partly a response to the stalled SCO. This initiative also represented a Chinese response to alternative ideas for economic cooperation in Eurasia, namely, the Eurasian Economic Union.[9]

Japanese involvement in this region was also especially significant in supporting the early nationhood of CA states in the early 1990s. As discussed in previous chapters, this support was framed within the Eurasian (Silk Road) Diplomacy championed by Prime Minister (PM) Hashimoto Ryutaro in 1997.[10] It was subsequently supported by Prime Ministers Obuchi Keizo, Mori Yoshiro and Koizumi Junichiro. By the early 2000s, Japan became one of the largest, if not the largest, official development assistance (ODA) providers to the CA countries. The Japanese government initiated the establishment of a Central Asia plus Japan dialogue in 2004, which aimed not only to establish a communications channel between Japan and CA states but also, and importantly, to encourage discussions of various regional problems by regional states. In this sense, the Japanese initiative has been one of the most significant decolonizing initiatives (offering alternatives to Russian and possibly Chinese funding and infrastructure, which were formerly routed through Russia) undertaken by non-regional powers in CA because it offered not only a forum for such discussions but also practical financial support from Japan toward projects on which more than two CA countries can reach a mutually acceptable agreement. To emphasize the importance of CA to Japan, then Japanese PM Koizumi visited Uzbekistan and Kazakhstan in 2006. A decade later, in October of 2015, PM Abe Shinzo visited all five CA states. In later years, PM Abe also promoted new initiatives such as High Quality Infrastructure[11] and the Free and Open Indo-Pacific Strategy to further advance Japanese corporate interests internationally and in the CA region.[12]

The South Korean presence in CA, and in Uzbekistan in particular, demonstrates some additional features to those seen in the Chinese and the Japanese policies. Like Japan, South Korea has demonstrated a desire for its own region-building scheme along the lines of Korea plus Central Asia, which to some extent was influenced by learning about the experiences of the Central Asia plus Japan initiative. However, some features distinguish Korea from China and Japan. The first is the presence of the large Korean diaspora, analyzed in detail in Chapter 8 of this volume. The second feature manifest in the South Korean penetration of CA is the fact that South Korean private/corporate interests are visibly more active and flexible when compared to public institutions and governmental agencies. By the time the South Korean government properly framed its initiative, roughly 15 years after the collapse of the USSR, South Korean Daewoo, Samsung, LG, Daewoo Unitel (a communications company), Kabool Textiles (a cotton processing and textile production company) and many other brands had thriving businesses in car manufacturing, textile processing and electronics assembly, most notably in Uzbekistan and Kazakhstan. Thus, the successes of these enterprises have pulled the Korean government's larger involvement into CA in a spill-over effect. Essentially, the successes of individual Korean enterprises sent a message to the Korean government that CA, and Uzbekistan in particular, is an area where Korean (public and corporate) interests can be served with great potential. This process has led to the Joint Declaration on Strategic Partnership with Uzbekistan in 2006 as well as other agreements opening new frontiers, including extraction agreements with the Korea National Oil Corporation (KNOC) in 2006; Uzbekneftegaz's granting to KNOC exclusive exploration rights to the Chust-Pap, Namangan-Terachi and Surgil gas fields in 2008; and purchases of uranium by the Korea Electric Power Corporation (KEPCO) in 2008, to name a few. Navoi airport has become another infrastructure project, with the Hanjin Group establishing and developing a Navoi logistical hub with a certain degree of success.

Thus, by the time South Korea's Silk Road Diplomacy was announced in 2009, South Korea's economic presence in CA, particularly in Kazakhstan and Uzbekistan, was significant in terms of ODA assistance (drawn by corporate successes of the early 1990s and 2000s), direct investments and human resource development. The Korea International Cooperation Agency (KOICA) had been allocating significant funds for various human resource development projects that were considered not only to be contributing to Uzbekistan but also to be preparing human resources needed for the Korean presence in CA and for South Koreans domestically. President Lee Myung-bak again visited CA in 2011 and took part in the opening of several enterprises with South Korean capital. This visit was followed by the initiation of a South Korea plus Central Asia dialogue. In this sense, in contrast with China and Japan, Korean region building did not start from political initiatives but instead grew out of economic and social engagement with the region.

The third feature that distinguishes Korea's engagement from that of China and Japan is the aspect of people-to-people communication. Here again, one can distinguish among these three cases. In the case of China, the access of the CA population to China is extremely limited, with visas very difficult to obtain, despite all the rhetoric of friendship between China and CA states. This limitation again shows that China is not interested in promoting people-to-people communications because it sees more threats than benefits from such contacts (namely, support for repressed ethnic Turkic groups such as Uighurs within China). Japan maintains a fair amount of people-to-people contacts with many scholarships extended to CA students and, in recent years, has eased visa regimes to allow those satisfying certain criteria to visit Japan for social reasons.

Uzbekistan, along with other CA republics, has abolished a visa requirement for short-term visits for Japanese tourists. South Korea, however, is perhaps the most advanced in these policies, as it not only issues visitation permits and visas for Uzbeks but also, importantly, attracts abundant human resources into the Korean labor market as described in Chapter 8. This demonstrates that the level of Uzbekistan's connection with and penetration by South Korea is not limited to corporate contacts only but extends to social spheres to a degree that cannot be compared to China and Japan.

As clearly outlined above, these East Asian countries have significant influence in the CA region and are increasingly diversifying their areas of interest. China ranks the highest in trade with Uzbekistan, while Korea is among the top economic partners, as seen in Figure 5.1. Japan is one of the largest ODA providers to CA, and to Uzbekistan in particular.

Despite the significant influence these states have, little if any literature focuses on the nuances of their influences and the features of their involvement rather than solely on the perceived rivalry among them. In addition, few details of their projects are provided, with no attempt to evaluate their performance in the CA region from a comparative perspective.

To fill this gap, this chapter aims to shed light on the place and role of CA states and Asian powerhouses (such as China, Japan and Korea) in one another's policies. To do so, this chapter analyzes the economic cooperation road maps agreed upon between governments of CA states and China, Japan and South Korea in 2017. Because coverage of all CA states is logistically difficult within this chapter, Uzbekistan has been selected as the case for CA states, and the cooperation road maps between Uzbekistan and its East Asian counterparts are this paper's main analytical focus.

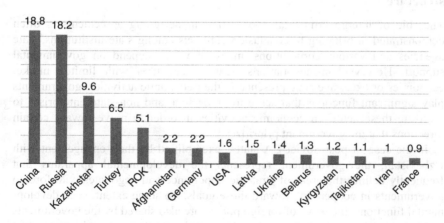

Figure 5.1 Uzbekistan's main trading partners and trade volumes, 2017.

Source: Vneshne-torgovyi balans Uzbekistana sostavil 17.8 milliarda [External Trade Balance of Uzbekistan Reached US$17.8 billion], *Gazeta*, August 14, 2018, accessed August 14, 2018, www.gazeta.uz/ru/2018/08/13/foreign-trade/.

This chapter raises the following research questions. What are the areas of interest for China, Japan and Korea in their relations with CA states? What are the patterns of agenda setting in establishing intergovernmental cooperation? What are the particular projects that these states initiate? What are the objectives of projects initiated within these areas of interest? How competitive or complementary are these projects of China, Japan and Korea?

Why is it all about the road maps?

The analysis of the economic cooperation road maps – sets of step-by-step plans for cooperation – is chosen as this chapter's main methodological tool for two reasons. First, while discourses on the intentions of various powers in engaging CA states have been analyzed on multiple occasions, few studies, if any, consider the particular projects these states plan, or analyze their reasoning and implementation. Second, while speeches and statements of presidents, foreign ministers and policy officials inform our understanding of the relations between CA and East Asian states, the road maps of cooperation demonstrate how these politically articulated intentions materialize in the practical realm. That is not to say that road maps are necessarily realizable plans. However, they are the most tangible plans that are closest to the practical outcomes of governments' articulated intentions and discourses. In this sense, this study attempts to understand the practical nuances of the engagement of China, Japan and Korea in this region by analyzing plan outlines within road maps worked out by intergovernmental committees of related states.

The economic cooperation road maps and the agenda-setting structure

The role of intergovernmental cooperation in designing agreements is often demonstrated in bringing to conclusion contracts among state institutions, state agencies and various corporations, many of which depend on governmental support. However, in negotiations between countries with limited market economies or excessive state presence in their economic activities, governments play significant functions that serve to guide state and non-state enterprises to motivate these actors in economic activities in order to move toward certain directions that the government prioritizes.[13]

In the case of post-Soviet CA states, exemplified by their engagement with Uzbekistan, the government represents the developmental apparatus that frequently defines the areas of strategic importance and negotiates with foreign governments in cooperating toward those goals. To some extent, such developmental functions, inclusive of foreign policy, are also shared by the governments of China, Japan and South Korea, making it easier for these countries to conduct their negotiations.

In terms of the negotiation structure between China, Japan, South Korea and Uzbekistan, they normally begin by establishing each state's objectives and goals as defined in their domestic developmental goals and programs. Each of these is communicated at different levels, but most conventionally through the channels of the ministries of foreign affairs. Often, such communications intensify when approximate dates for visits of heads of state, governments or foreign ministers are decided. Once those domestic programs and goals are mutually articulated and duly recognized, the possible areas of cooperation are distilled from those programs. Most typically for the countries covered in this chapter, the areas of trade, transportation infrastructure development, energy resource exploitation, innovation and technologies and security are considered to be of primary importance. The degree of importance of each area fluctuates depending on the country. For instance, in cooperation with China, infrastructure development and security feature prominently, while in cooperation with Japan and South Korea human development, technological innovation and modernization of infrastructure receive higher attention and importance. Once the areas of cooperation are defined, each government involved designates the main actor responsible for promoting cooperation within this area. In the case of Uzbekistan and some other post-Soviet states, the degree of centralization of governmental functions is very significant and often results in a situation where a single ministry (for instance the Ministry of Economy) is the main actor in negotiating cooperation in several areas (such as infrastructure, transportation or energy resources). In the cases of China, Japan and South Korea, the situation drastically differs from one case to another. For instance, in the case of China the degree of centralization is somewhat close to CA states, so the same ministry is often responsible for promoting cooperation in the same area. In the cases of Japan and South Korea the situation is very different, largely reflecting the degree of economic liberalization and the central government's decentralization, as described in the country-specific parts of this study. Once the main actors are defined by each government, these actors determine the most appropriate agencies and actors to assist them in promoting this area of cooperation.

This process prepares the stage for proposals to be forwarded by all actors within the area of cooperation to be included in the proposal for action map. Normally, each side considers proposals for each area and then decides whether they want to forward these proposals to their counterparts. Initial discussions can be held through channels provided by ministries of foreign affairs and unofficially signaled to counterparts. In certain other cases, the actors of a potential cooperation scheme visit the country or site of possible cooperation to assess both the potential and the challenges for developing a successful cooperation scheme. This action is also often needed when certain public or private enterprises design their business plans, requiring a detailed assessment of local conditions, needed investments and possible revenues to be generated from the project. Such potential plans are preliminarily signaled to the counterpart's ministry of foreign affairs (MOFA), simplifying the process of facilitating the visits and gaining access to the data required for risk calculation. In this sense, the roles played by the governments in such cooperation schemes might not be necessarily to invest public funds into these projects but rather to provide a secure environment for private enterprises to enter the markets of countries they traditionally consider to be risk prone, such as Uzbekistan. Once the projects are offered as proposals, a meeting is held between the intergovernmental committee on cooperation and its subcommittees focusing on particular areas and consisting of representatives of both sides.

When the interests of the different sides do not match, they move on to projects that are of greater mutual interest. In this sense, the establishment of the committee and the respective subcommittees creates a channel of communication and a sort of bargaining table that is open throughout the year on an ad hoc basis. This committee and its subcommittees are also a good way to signal certain policy and priority changes for each country. They help prevent miscommunications at the political level and provide coordination capacity for enterprise activity.

While the Chinese government is well prepared and experienced in playing such leading and often developmental roles for its corporations and enterprises, the case of Japan demonstrates that the Japanese government is rather hesitant to play an active role in facilitating private enterprise entry into the CA region, primarily because Japan has a completely liberalized and free market economy. In such a structure, Japan's government (and the Japanese MOFA in particular) is reluctant to play a role in singling out a particular enterprise and promoting its interests, which might be interpreted as governmental interference in economic activity. Such a situation, however, does not necessarily represent a structural

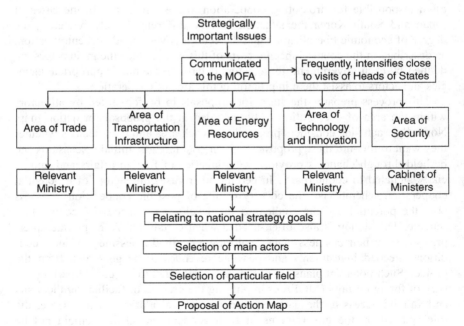

Figure 5.2 The structure of the initiation of cooperation for Uzbekistan.

problem, and there continues to be an opportunity for the Japanese government to promote its private enterprises in the CA region without being accused of interference, as detailed in the section on the Japanese road maps below.

The Korean case is somewhat different. Although in the Chinese and the Japanese cases the central government is frequently the engine for encouraging direct investments into CA economies, in the Korean case private enterprises are far more active in promoting cooperation, while the government plays reactive roles with respect to such entrepreneurial activities. The Korean government does not play the pivotal role in initiating entrepreneurial activities, but it is often pulled into playing a more prominent role in the region where Korean enterprises have already built a significant economic presence. In addition, the spill-over effect occurs to a certain degree in Korean involvement in this region when successful projects by certain enterprises encourage development of similar ones in other

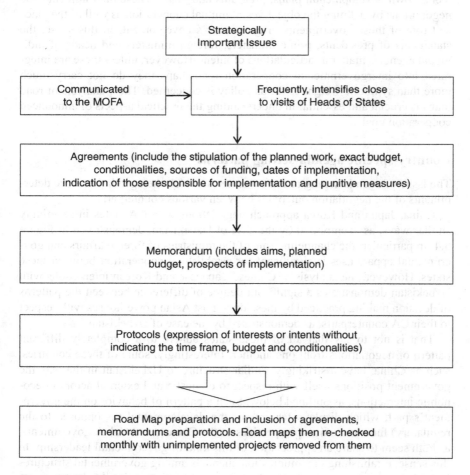

Figure 5.3 The structure of agreements.

areas that are predominantly private-interest driven. Such a spill-over effect is not necessarily observed in the cases of Chinese and Japanese private participation.[14]

Once the proposals for each of the main actors responsible for cooperation in these areas are considered and analyzed, only those deemed to potentially produce tangible short-to-mid-term outcomes are included in the proposals for each area of cooperation. The proposals are then grouped into framework agreements, contracts, and memoranda that constitute the backbone of intergovernmental cooperation road maps (see Figure 5.3). In this sense, the intergovernmental cooperation road maps are sets of plans, agreements and memoranda that document commonly shared norms, approaches and objectives between Uzbekistan on the one hand and China, Japan or Korea on the other.

These economic cooperation road maps often reflect not only the governments' intentions and goals but also, and importantly, the negotiating governments' own developmental plans, programs and goals. These then influence the negotiations by defining the objectives of mutual cooperation as well as the place and role of these governments in each other's development. In this sense, the statements of presidents, prime ministers, foreign ministers and heads of individual agencies matter as articulations of intent. However, unless these are integrated into intergovernmental cooperation road maps they do not carry much more than symbolic weight as far as policy is concerned. Thus, analysis of road maps is crucial and essential in understanding the practical aspects of announced cooperation goals.

Country-specific agenda-setting patterns

The issue of agenda setting and the manner in which the agenda is set are determinants of the negotiation outcomes between various counterparts.

China, Japan and Korea approach negotiations with CA states in seemingly similar ways, as exemplified by the case of Uzbekistan, demonstrated in Figure 5.4. In particular, the channeling role of the ministries of foreign affairs and governmental apparatuses is significant for establishing cooperation between these states. However, the analysis of Chinese, Japanese and Korean interactions with Uzbekistan demonstrates a significant degree of difference between the patterns of decision making practiced by these three East Asian powerhouses with respect to their CA counterparts, as demonstrated by the case of Uzbekistan.

That is not to say that all of these states adhere to a completely different pattern of negotiations from one another. Interestingly, some of these countries, such as China, have a strikingly similar structure to Uzbekistan in the way the government positions itself with respect to domestic and external actors in economic interactions, as outlined below. Such a pattern of behavior on the government's part, which displays features of the developmental (as opposed to the regulatory) function, unites perspectives of the Chinese and Uzbek governments, as both seem to share an appreciation of this kind of governmental leadership. In this sense, establishing communication channels among governmental structures is detrimental to establishing effective inter-state economic ties.[15]

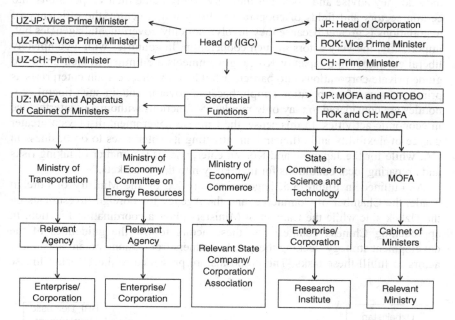

Figure 5.4 The structure of Intergovernmental Committees and roles played by various actors.

Note
IGC = intergovernmental committee; UZ = Uzbekistan; JP = Japan; ROK = Korea; CH = China.

First, domestic signposting documents are similar and guide the agenda setting on both the Chinese and the Uzbek sides. In the Uzbek case, this is the Development Strategy 2017–2021, which has been analyzed elsewhere.[16] This document's significance is that it sets important goals and objectives for Uzbekistan's economic development, which serve as critical signposts when approaching foreign counterparts. Unless these goals and objectives are met, Uzbekistan does not display a strong desire to enter into agreements. On the Chinese side, the Ministry of Foreign Affairs, the National Development and Reform Commission, and the Ministry of Commerce drafted an action plan in 2015 outlining policy coordination, connectivity, unimpeded trade, financial cooperation, and people-to-people bonds as the primary principles of Chinese engagement in other regions, including CA.[17] The same type of commitment has been displayed by the Uzbek governmental bureaucracy, which prepared its own road map emphasizing cooperation at the governmental level to facilitate trade.[18]

Second, the similarities in patterns of governance between China and Uzbekistan can be cited as another factor significantly affecting cooperation's success, which for the purposes of this chapter, is defined as the number of agreements signed within the framework of the economic cooperation road maps. In particular, in the case of the China and Uzbekistan governments, not only do they advise and assist but they essentially guide their corporations into certain sectors and fields of cooperation. In such a structure, the efficiency of negotiations is greater because they involve not only governmental agencies but, importantly, the very actors in such cooperation. In countries with a completely liberal economy, like Japan or Korea, governments are limited in their ability to guide private corporations into particular fields or to select certain enterprises as primary actors of cooperation simply because governments in such liberal economies are considered to play only regulatory functions without any interference in economic activities. In this sense, the Chinese government displays a greater degree of flexibility and efficiency in attracting its enterprises to do business in CA, while for the Japanese and Korean enterprises it is a matter of taking risks and preparing local conditions for their entry into those markets.[19]

As outlined in Figure 5.5, negotiations are led by the Ministry of Foreign Trade, the Chamber of Commerce, and the Agency for Foreign Investments on the Uzbek side while the Cabinet of Ministers plays a coordinating function. In approaching Chinese counterparts, these actors follow the guidelines of the development strategy, seeking the best-fit partners from among potential Chinese actors to fulfill these tasks. The proposals are presented to the Uzbek–Chinese

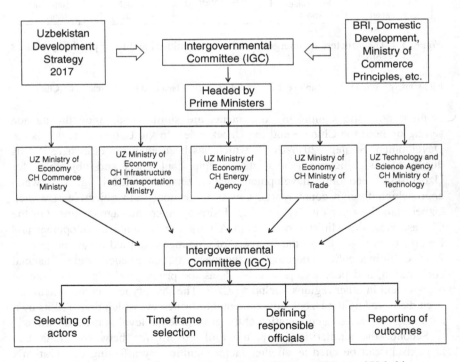

Figure 5.5   The actors in and patterns of negotiations between China and Uzbekistan.

Note
UZ = Uzbekistan; CH = China.

Intergovernmental Committee for Cooperation, co-headed by the Uzbek deputy prime minister and the secretary of the Chinese Communist Party's Political-Judicial Committee.[20]

The intergovernmental committee's work is divided among several subcommittees, such as those on trade and economic relations (coordinated by the Uzbek Ministry of Foreign Trade and the Chinese Ministry of Commerce), cooperation in energy (coordinated by the Uzbek Ministry of the Economy and the PRC National Energy Administration), transportation (coordinated by the Uzbek Ministry of Foreign Trade and the PRC Ministry of Transport), technical and scientific cooperation (coordinated by the Uzbek Committee on Coordination of Science and Technology and the PRC Ministry of Science and Technology), cultural and humanitarian cooperation (coordinated by the Uzbek Ministry of Culture and the PRC Ministry of Culture), and cooperation in security-related issues (coordinated by the Uzbek Ministry of Foreign Affairs and the PRC Ministry of Foreign Affairs).[21]

These committees meet at least once a year, with the issues to be discussed communicated through the channels of the coordinating institutions well ahead of the meetings. If the needs of Uzbekistan can be met by a Chinese enterprise, the Chinese coordinating agency frequently serves as the communication line to establish the contacts and invite the potential investors from China to discuss the project. Additionally, to facilitate connections between businesses, several ministries (for instance, the Uzbek Foreign Trade Ministry and the PRC Ministry of Commerce) have signed memoranda to organize producer exhibitions, which have eventually resulted in the visit of small- and medium-sized enterprise representatives from Tianjin to the Jizzakh and Sirdarya free economic areas and the decision to allocate certain areas in Jizzakh exclusively for entrepreneurs from that region of China.[22]

However, this coordination capacity is only made possible through a certain degree of structural similarity in the relations between the governments and businesses, because these governments still play a much greater role than they would in countries with Western liberal market economies. In addition, once an agreement is reached, the government of Uzbekistan still exercises significant control over the economic activities of the enterprises, which is problematic for companies from other countries but acceptable to Chinese corporations due to the similarity of governmental controls in China. On the Uzbek side, a separate committee is created to ensure the agreements created for each partner country are implemented. For instance, aside from China, the latest development in this regard was the creation by Uzbekistan of a national committee to ensure the implementation of agreements with the United States.[23] The committee's work is stipulated by road maps for cooperation with each country, and committee heads present reports on progress and the relevant directions of the road maps to the president between the fifth and the tenth of each month.

As a result of this work, the focus of the May 2017 visit of President Shavkat Mirziyoyev to China was obviously on the promotion of economic ties between the two countries based on the "Shanghai spirit" bargaining strategy.[24] Even before this visit, Uzbekistan and China enjoyed a fair level of economic cooperation, as seen in the list of investment projects underway prior to 2017 announced by the Uzbek Ministry of the Economy.[25] The visit marked the signing of one of the most ambitious packages of agreements, including 11 inter-governmental agreements, one intermunicipal agreement and a package of economic contracts worth US$22.8 billion.[26] It remains to be seen how many of these projects will reach their declared outcomes, and there is no comparable data to indicate the general ratio of implementation of these projects. However, as has been indicated to the author by the Uzbek government official anonymously interviewed in 2017, the fact of inclusion into the road map puts considerable pressure on government officials to do their utmost to ensure their implementation. Institutionally, the degree of implementation of these projects is checked at the governmental meetings held monthly, which when deemed necessary make needed corrections to ensure implementation of these plans.

As described in the sections below, the most significant areas in which Chinese road maps envision cooperation are manufacturing, resource-related investments and infrastructure development.

In contrast to the Chinese agenda-setting pattern, Japan's approach represents a significantly different government–business relationship. While the Chinese–Uzbek inter-state committee is primarily composed of government members, government-affiliated agencies and state-run corporations as well as financial institutions, the Japan–Uzbek inter-state cooperation committee primarily aims to unite the prospective market actors. Thus, it positions itself as an institution representing the interests of a wider spectrum of actors, going beyond governmental institutions. The main idea of governmental participation in these negotiations is that the Ministry of Foreign Affairs and other related ministries merely play the role of facilitators of the dialogue between the private economic entities and humanitarian organizations. Such a prominent role of the governments sometimes leads to abuses of authority on the part of both the host and the

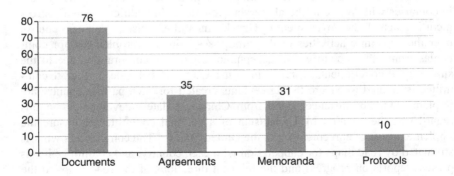

Figure 5.6 Constituent elements of the China–Uzbek road map.

Source: "Uzbekistan i Kitai Podpisali Ryad dokumentov" [Uzbekistan and China Signed a Range of Documents], *Review.uz*, May 13, 2017, accessed September 23, 2017, www.review.uz/novosti-main/item/11214-uzbekistan-i-kitaj-podpisali-ryaddokumentov-spisok.

investors, such as the situation when some companies (in particular Nihon Koutsu Gijyutsu [Japan Transportation Consultants])[27] were caught paying bribes to Uzbek (and to Vietnamese and other states') officials in the process of ODA implementation, leading governments to place safeguards against such situations.[28]

Although the Japanese government plays an important role in arranging a proper platform for dialogue, neither the Japanese foreign ministry nor any other state body aims to adopt a developmental role (persuading the Japanese participants, assisting in selecting the companies, taking part in negotiating better treatment, etc.) or to lead the process. In addition, the Japanese government does not adopt responsibility for defining the areas of most pressing concern for its businesses, but instead allows corporate interests to lead the discussion. In such a situation, and in Uzbekistan in particular, governmental desire alone does not represent sufficient support for certain corporate interests to get involved in CA.

As if to reflect on this difference with China, the Inter-State Committee on Economic Cooperation between Japan and Uzbekistan is composed not of public officials but largely of representatives from the commercial sector. These include representatives of various corporations on the Japanese side, with the Japan Association for Trade with Russia & NIS (ROTOBO, NIS signifying the New Independent States of CA) playing the role of coordinator for these activities. This representation also demonstrates the structural mismatch between Japanese and Uzbek expectations. On the Japanese side, an expectation exists that the corporate community will express a desire to enter the Uzbek market once it gains confidence through information-sharing meetings and an increase in personal contacts. Therefore, the main actor in such interactions is the corporate community. On the Uzbek side however, an expectation exists that the Japanese government needs to play a more active part not only in setting the stage for information exchange but, more importantly, in encouraging particular Japanese corporations to enter the Uzbek market. Such an expectation is well understood among Japanese policymakers. However, given the limitations of free economic enterprises, the Japanese government limits its role only to information gathering and providing basic support to its corporations. Although the Japanese government's position fits well with the principles of a liberal economic system, such a "birdwatching" stance does not seem to bring any tangible outcomes, since the other East Asian governments of China and Korea do not hesitate to openly promote the interests of particular corporations. The coordination role of ROTOBO has also been criticized both by Uzbekistan and by its Kazakh counterparts on several occasions. The reason for such criticism emanates from the fact that CA governments are represented at such meetings by representatives of ministries and state agencies that are in a position to make policy and practical decisions. At the same time, ROTOBO is an organization that is neither in a position to propose a particular policy change nor able to implement any policy decisions. It is not a part of the executive body, and most of its views and perceptions are of a consultative nature, which has little or no relevance to the policy field. In addition, the Inter-State Economic Committee is mismatched: on

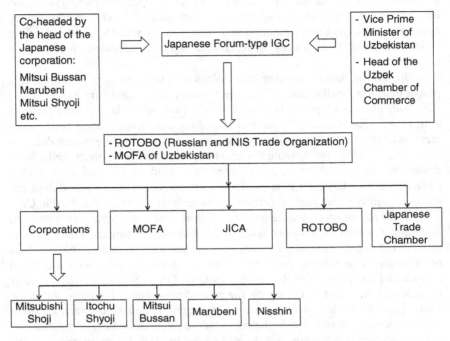

Figure 5.7   Japan–Uzbek Economic Cooperation Committee composition.

Note
IGC = intergovernmental committee.

the Japanese side, it is jointly led by a representative of a Japanese corporation on a rotational basis while on the Uzbek side it is supervised by the prime minister or the deputy prime minister.

Although the logic of this structure is that the committee becomes a meeting place between Japanese businesses and Uzbek bureaucracy, in practice these meetings do not result in expected outcomes because these actors operate with different objectives. While the Uzbek bureaucracy operates on behalf of the government, the Japanese side aims at airing corporate voices, which are not connected to policy. In many cases, these committees turn into forum-type meetings that articulate many desirable objectives but produce very few tangible outcomes.

As is often the case, entrepreneurs frequently express desires to enter certain markets and areas, including in Uzbekistan, but unless the conditions are prepared in terms of the legal and financial infrastructure, the Japanese corporate community displays a significant degree of hesitance in entering such markets, despite strong governmental support in facilitating such entry. This is perhaps the greatest difference in Japanese and Chinese corporate behaviors, because Chinese corporations seek to compensate for the absence of legal and financial infrastructures through agreements between the governments that provide additional guarantees to the corporations entering Uzbekistan.

With regard to Japanese and Korean corporate behavior, certain similarities exist between the two. However, Korean entrepreneurs seek to utilize the opportunities received in the negotiations during governments' official visits, and if those agreements do not materialize well many Korean companies immediately withdraw without hesitation. Therefore, the Chinese approach represents "high risk compensated by governmental guarantees", while Korean corporate behavior can be summarized as "high risk, high return versus low return, fast retreat" and the Japanese behavior is closer to the model of "no risk, low/no return".

The composition of these countries' inter-state economic cooperation committees with Uzbekistan significantly influences the outcomes of deliberations. In line with the above explanations, intergovernmental economic cooperation plans between China and Uzbekistan are heavily dominated by the projects implemented by private enterprises in the fields that remain of high interest to both the Chinese and Uzbek governments. Therefore, one can observe the invisible hand of both governments in guiding their enterprises into the fields they consider important for state development. In the case of the Japan–Uzbek commission, one can observe the dominance of intergovernmental agreements and

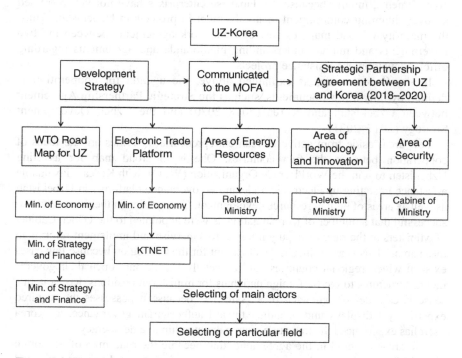

Figure 5.8 Korea–Uzbek Inter-State Economic Cooperation Committee composition.
Note
UZ = Uzbekistan.

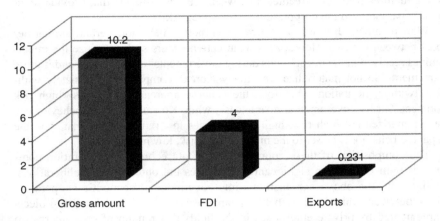

Figure 5.9 Korea–Uzbek road map agreements and composition (US$ billions).

Note
The figures are for a total of 67 documents; FDI = foreign direct investment.

commitments, mainly because the Japanese enterprises have not yet expressed an overwhelming commitment to involvement in projects in Uzbekistan. Thus, the majority of road maps consist of framework agreements between the two governments and mutual understanding memoranda and agreements regarding official development assistance projects.

In the Korean case, as explained above, the agenda for cooperation is defined by only a few documents, such as the Strategic Partnership Agreement between Uzbekistan and Korea (2018–2020) and the Uzbek Development Strategy (2017–2021).

Few areas aside from active Korean entrepreneurial activities are the focus of cooperation between the two countries. One is the road map for assisting Uzbekistan to join the World Trade Organization (WTO), with Korea's invaluable assistance in setting up electronic trade and e-governance platforms in Uzbekistan as well as areas of human resource development. Cooperation in these areas was so successful that a number of Korean nationals were appointed to the Uzbek Cabinet of Ministers at the rank of deputy minister to supervise and implement reforms in these areas. This is a significant development for the CA region because few cases exist in which regional countries, and especially Uzbekistan, open their governmental structures to recruit foreign nationals for ministerial positions. Korea in this sense is considered somewhat of a "safe bet" because it possesses the required expertise yet displays understanding toward authoritarian governance, as Korea itself has experienced the transition from a dictatorship to a democracy.

The areas covered in the agreements that became the road map of economic cooperation are dispersed across very wide areas of coverage. Each area outlined in Figure 5.10 is composed of a number of agreements, each of which aims at smaller objectives and is often implemented by a different actor. Therefore, the

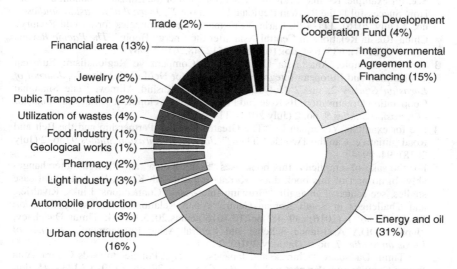

Figure 5.10 The areas and projects of cooperation included in the Korea–Uzbek road map of cooperation.

road map of cooperation with Korea in its essence is denser and more complicated than the Japanese road map. It might not be as significant in terms of the overall amount covered by its projects when compared to the one with China, but in terms of project diversity and actor multiplicity, it surpasses the latter.

Conclusions

Economic cooperation road maps presented in this chapter merely represent the intentions of the governments and other non-governmental organizations to pursue certain goals and objectives. In this sense, this chapter treats these road maps as a type of political narrative. Thus, the fact that these road maps have been agreed upon does not necessarily imply that they will be implemented. However, they still represent clearly formulated documents with actors, budgets and timeframe definitions that can be treated as generating certain political messages.

Accordingly, the economic cooperation road maps are indicative of the approaches and goals of China, Japan and Korea in this part of the world. They indicate the expectations of cooperating with CA states, exemplified by the case of Uzbekistan.

Notes

1 Timur Dadabaev, "The Last Asian Frontier? A Comparison of the Economic Cooperation Agendas of China, Japan, and South Korea with Uzbekistan", *The Asan Forum* (2018): 1–12.

2 See, for example, Ksenia Kirkham, "The Formation of the Eurasian Economic Union: How Successful Is the Russian Regional Hegemony?", *Journal of Eurasian Studies* 7, no. 2 (July 2016): 111–128; Carla P Freeman, "New Strategies for an Old Rivalry? China–Russia Relations in Central Asia after the Energy Boom", *The Pacific Review* 31, no. 5 (2018): 635–654, doi:10.1080/09512748.2017.1398775.

3 See, for example, Anastassia Obydenkova, "Comparative Regionalism: Eurasian Cooperation and European Integration. The Case for Neofunctionalism?", *Journal of Eurasian Studies* 2, no. 2 (July 2011): 87–102; Rashid Alimov, "The Shanghai Cooperation Organisation: Its Role and Place in the Development of Eurasia", *Journal of Eurasian Studies* 9, no. 2 (July 2018): 114–124.

4 See for example, Yongquan Li, "The Greater Eurasian Partnership and the Belt and Road Initiative: Can the Two Be Linked?", *Journal of Eurasian Studies* 9, no. 2 (July 2018): 94–99.

5 For the sake of simplicity, this book uses "South Korea" and "Korea" interchangeably. In no part of this book does "Korea" imply North Korea. For individual case studies, see Matteo Fumagalli, "Growing Inter-Asian Connections: Links, Rivalries, and Challenges in South Korean–Central Asian Relations", *Journal of Eurasian Studies* 7, no. 1 (2016): 39–48, doi:10.1016/j.euras.2015.10.004; Timur Dadabaev, "Japan's ODA Assistance Scheme and Central Asian Engagement", *Journal of Eurasian Studies* 7, no. 1 (January 2016): 24–38.

6 See Timur Dadabaev, "Chinese and Japanese Foreign Policies towards Central Asia from a Comparative Perspective", *The Pacific Review* 27, no. 1 (2014): 123–145, doi: 10.1080/09512748.2013.870223; Timur Dadabaev, "Discourses of Rivalry or Rivalry of Discourses: Discursive Strategies of China and Japan in Central Asia", *The Pacific Review* (2018, print forthcoming 2019), doi:10.1080/09512748.2018.1539026.

7 See Dadabaev, "The Last Asian Frontier?".

8 Timur Dadabaev, "Shanghai Cooperation Organization (SCO) Regional Identity Formation from the Perspective of the Central Asia States", *Journal of Contemporary China* 23, no. 85 (2014): 102–118, doi:10.1080/10670564.2013.809982; Dadabaev, "Chinese and Japanese Foreign Policies".

9 Timur Dadabaev, "Engagement and Contestation: The Entangled Imagery of the Silk Road", *Cambridge Journal of Eurasian Studies* 2018, no. 2, doi:10.22261/cjes. q4giv6.

10 For details, see Timur Dadabaev, "Central Asia: Japan's New 'Old' Frontier", *Asia Pacific Issues* 136 (February 2019): 1–12.

11 Ministry of Foreign Affairs (MOFA) of Japan, Presentation, www.mofa.go.jp/ files/000117998.pdf, accessed September 25, 2018.

12 For details of the Indo-Pacific Strategy see MOFA of Japan, "Priority Policy for Development Cooperation 2017", www.mofa.go.jp/files/000259285.pdf.

13 For example, see Decree of the President of Uzbekistan on Reconfirming "Agreement between Republic of Uzbekistan and People's Republic of China on Facilitating Smooth International Road Transportation", PP-3143, July 20, 2017, in Russian at http://nrm.uz/contentf?doc=508717_postanovlenie_prezidenta_respubliki_uzbekistan_ot_ 20_07_2017_g_n_pp-3143_ob_utverjdenii_mejdunarodnogo_dogovora&products=1_ vse_zakonodatelstvo_uzbekistana, accessed September 23, 2017.

14 For examples of this, see Timur Dadabaev, "Japanese and Chinese Infrastructure Development Strategies in Central Asia", *Japanese Journal of Political Science* 19, no. 3 (2018): 542–561, doi:10.1017/S1468109918000178.

15 For details, see Dadabaev, "The Last Asian Frontier?".

16 Uzbekistan's Development Strategy 2017–2021, accessed September 25, 2018, http:// old.lex.uz/pages/getpage.aspx?lact_id=3107042; Timur Dadabaev, "Uzbekistan as Central Asian Game Changer? Uzbekistan's Foreign Policy Construction in the Post-Karimov Era", *Asian Journal of Comparative Politics* 4, no. 2 (2018, print forthcoming 2019): 162–175, doi:10.1177/2057891118775289.

17 Sarah Lain, "Russia and China: Cooperation and Competition in Central Asia", in *Chinese Foreign Policy under Xi*, ed. Tiang Boon Hoo (London: Routledge, 2017), 74–95.

18 For instance, Uzbek Foreign Trade and the PRC Ministry of Commerce signed memoranda to facilitate the organization of producer exhibitions in May 2017. See Dadabaev, "The Last Asian Frontier?".

19 For instance, in a private conversation in September 2017 in Tashkent, a Japanese embassy official described such an Uzbek–Japanese cooperation structure as impossible due to a regulatory function of the Japanese government, which prevents it from forcing Japanese corporations into foreign cooperation schemes.

20 See "Agreement between Governments of the Republic of Uzbekistan and the People's Republic of China on Creation of Uzbek-Chinese Intergovernmental Committee for Cooperation, Beijing", 2011, www.lex.uz/pages/GetAct.aspx?lact_id=1986611.

21 "Agreement between Governments of the Republic of Uzbekistan and the People's Republic of China". See the appendices to the agreement.

22 See www.jahonnews.uz/en/ekonomika/316/36956/.

23 Decree of the President of Uzbekistan, accessed September 23, 2017, PP-3293, www. norma.uz/raznoe/postanovlenie_prezidenta_respubliki_uzbekistan_ot_27_09_2017_g_ pp-3293.

24 "Shanghai spirit" refers to a norm that connects the issues for cooperation prioritized by both countries without seeking unilateral gains. This norm also implies the importance of mutual sacrifices and compromises for mutual gain.

25 Ministry of Economy of Uzbekistan, https://mineconomy.uz/ru/node/1091, accessed September 23, 2017.

26 This amount includes both Chinese and Uzbek shares in the deals signed. Thus, the pure Chinese contribution/investment in the deals is roughly US$10 billion; the rest accounts for Uzbekistan's contributions, including monetary contributions as well as the costs of land and infrastructure development.

27 MOFA of Japan, www.mofa.go.jp/mofaj/press/release/press4_001142.html.

28 MOFA of Japan, www.mofa.go.jp/mofaj/files/000047885.pdf.

6 The Chinese economic "offensive" in post-Karimov Uzbekistan

The strengthening of Chinese influence in Central Asia (CA) has been a result of an overall increase in China's global influence. Indeed, the patterns of Chinese economic and political expansion in CA are not much different from those in other parts of the world. CA represents an area of vital interest for China for various reasons, including security concerns (such as separatism, terrorism and extremism); economic potential (CA is an energy resource–rich area with a vast market for Chinese goods); political importance (balancing the presence of other powers such as Russia and the Afghanistan-invested US); and infrastructure potential (CA could become a transportation hub for Eurasia). In this sense, the motivations behind the Chinese presence in this region are clearly outlined in Chinese official discourse as well as in the analyses of academicians.[1] However, in regard to the interests and motivations of CA states, the coverage is mainly limited to wishing to join the bandwagon. The general picture of CA states is usually drawn from the real or imagined attempts of these states to balance the powers of China, Russia, and the US (and other states); new Great Game rhetoric; or a purely utilitarian attitude toward China's vast financial resources and market for energy resources.[2] While Chinese motivations for engaging in CA are fairly well covered in the literature, few studies analyze how Chinese economic expansion fits the domestic plans and priorities of Central Asian states.[3] And few studies focus on the role and place of China (both its government and its corporate community) in the modernization of these societies. There is also little attempt to describe not only the outcome of Chinese–Central Asian cooperation but also the process of agenda setting for cooperation and how this agenda is then implemented.[4] Chinese and Central Asian policymakers often note that their cooperation has mutual benefits and that all sides involved are in a position to set and redefine the agenda for cooperation. However, in many cases, the commentaries on such cooperation schemes tend to emphasize the economic strength of the Chinese and their persistence in lobbying for economic benefits in CA, while Central Asian states are depicted as mere consumers of the Chinese economic, political and social agenda.[5]

Uzbekistan's relations with China are of special importance in this regard. As is frequently noted by scholars and practitioners, Uzbekistan is centrally located, resource rich (especially in rare metals and natural gas), and the most populous country in CA, with a population almost equal to the population of all other Central Asian states combined. Such advantageous central, resource and demo-graphic positioning alone suggests that any instability in this country may raise questions about the stability of CA in general and have a critical effect on Chinese plans for the Belt and Road Initiative and other schemes in this region.[6]

Above all, what attracted region watchers and experts to the foreign policy of Uzbekistan recently was the long-anticipated change of power after the death of Islam Karimov, the first president of Uzbekistan, in September 2016, and the appointment of Shavkat Mirziyoyev as interim president, and then his election to the presidency in December 2016.[7] While Karimov was famous for balancing the influence of all regional and non-regional states in his quest to solidify his hold on economic and political power, the change of leadership signified an opening up of the country to new engagement with the international community. At the same time, this opening up has raised many questions about the direction of Uzbekistan's foreign policy in the post-Karimov era. Some scholars predict a policy of distance from all powerful states, while others argue that Uzbekistan will follow a pro-Russian or, alternatively, a pro-Chinese foreign policy.[8]

Considering these issues, this chapter raises several questions: How has the change of leadership influenced the Uzbek–Chinese cooperation agenda? What are the areas that these states prioritize in engaging each other? What are the institutions and motivations for setting up the agenda for this cooperation?

By providing answers to these questions, this chapter first contributes to the understanding of the impact of Uzbekistan's change of power and its evolving relations with China. Second, it explains the motivations of strategically important (in terms of economy, demography and political capability) Central Asian states such as Uzbekistan in engaging China. Third, this chapter provides details of Chinese engagement in this region by outlining the latest project details. Although China's Belt and Road Initiative has received wide coverage, few details of the Chinese projects that constitute Belt and Road engagement (aside from infrastructure) and their impacts have been analyzed.[9]

To answer the questions listed above, this chapter first analyzes two major documents adopted after the change of leadership in Uzbekistan: the Develop-ment Strategy of Uzbekistan for 2017–2021 and the January 11, 2018[10] foreign policy speech of President Mirziyoyev. Both documents define priorities in Uzbekistan's development and the regional and non-regional states that Uzbekistan considers most important. Through this analysis, this chapter aims to identify the role of China (on par with other East Asian states) in Uzbekistan's emerging foreign policy in the post-Karimov era. Next, this chapter explains how cooperation between Uzbek and Chinese policymakers in the areas of Uzbek concern is made possible by describing how the intergovernmental bureaucracy works and the norms and practices of bureaucracy shared by both countries. The analysis then outlines several areas of cooperation, with a descrip-tion of the agreements achieved during the visit of President Mirziyoyev to China in May 2017, and how these agreements have been implemented. This part of the chapter aims to demonstrate that the agreements signed in Beijing focus almost exclusively on areas coinciding with those selected for the national development strategy of Uzbekistan. Although there are very important security interests shared by China and Uzbekistan, due to space limitations and the large number of studies addressing this topic, this chapter does not describe the history of the Shanghai Cooperation Organization's relations with Uzbekistan or the history of diplomatic relations that can be found elsewhere.[11] It focuses almost exclusively on explaining the development of various areas of cooperation under the new leadership from September 2016, to highlight the latest trends and developments in relations between the two countries.

Through all of its sections, this chapter aims to deliver the message that Uzbekistan is engaging China in a new domestic and international environment of opening up after decades of Karimov's rule. In particular, Mirziyoyev aims to change the strategy of Uzbekistan from reactionary (mainly reacting to changes in the international environment) toward proactive. In such a shift, China (in economic terms) is attributed an important role as a source of technology transfer, an investor and a market for Uzbek resources and goods. Russia, in turn, represents a vast market for agricultural products and certain industrial goods (cars, certain electronics), as well as a labor market. By modernizing its technological assets and the industrial sector of its economy, Uzbekistan aims to create an economic structure that will allow it to reduce the share of imports from other countries and, importantly, increase the share of exports, not only to China but also to other Central Asian states. Uzbekistan aims to take advantage of its abundant human resources and cheap labor to create an export-oriented economy with respect to neighboring countries that suffer from either energy resource–centered economic development or a lack of diversified manufacturing capacity. As described below, Chinese technology transfer might not offer the most advanced technology, but it is sufficient for Uzbekistan's goal of creating an export-oriented industrial base. In cases where more advanced technology is required, Uzbekistan chooses the Republic of Korea or Japan for technology transfer.[12] In this sense, the evolving foreign policy of Uzbekistan displays flexibility in choosing international partners depending on the country's need to support economic restructuring.

National development strategy of Uzbekistan and the role of China

President Karimov's foreign policy has been described in various ways. In the early 1990s, Uzbek foreign policy was regarded by many as emphasizing the importance of Russia and Turkey.[13] The expert community assumed that this trend was later reconsidered in light of perceived threats from Turkey[14] and a lack of commitment from Russia.[15] Others saw signs of Uzbekistan's pro-Western policy in the late 1990s and early 2000s. However, US criticism of Karimov's handling of the Andijan events of 2005[16] and of the wave of color revolutions welcomed by the US and Europe pushed Uzbekistan to the offensive. Uzbekistan has since been considered to be building closer economic and political ties with Russia, China and South Korea, while treating US and European initiatives as threatening to its sovereignty. These drastic shifts were described as reactionary and opportunistic. With intensification of the Russian Eurasian initiative and the Chinese Belt and Road Initiative in the later 2000s, experts described Uzbekistan as moving to balance Russian and Chinese influences by gradually improving its US ties to diversify its external partners.

When Mirziyoyev became president, Russian and other media sources speculated that Uzbekistan would move closer to Russia because of his perceived close personal ties to Russian elites.[17] However, Uzbekistan has since declared several principles for new foreign policy behavior that discursively adhere to the old policy but, in practice, demonstrate a drastic departure from the behavior of Uzbekistan under Karimov. First, Mirziyoyev chose a foreign policy direction centered on non-alignment with military and political blocs and non-intervention in the affairs of neighboring states.[18] This principle was reconfirmed in his first address to the joint session of the chambers of parliament in September 2016. Second, Uzbekistan announced that its main priorities lie in improving and maintaining its relations with neighboring CA states. This principle had consistently been stated in previous years, but there was little evidence of its being operationalized. Mirziyoyev demonstrated his commitment through 11 visits, two working visits and 15 phone calls with presidents of neighboring countries in his first year in office. This produced outcomes that had been difficult to achieve over the 25 years of independence of these states. Uzbekistan managed to find compromises with Kyrgyzstan over border- and water-related problems; with Tajikistan over transportation networks, air communications, and cultural exchange; and with Kazakhstan over the opening of Uzbek markets for imports and promoting Uzbek exports and eventual intensification of trade. Uzbekistan also announced that it aims to develop its economic relations to export domestically produced goods to neighboring countries and even to establish joint manufacturing plants for Uzbek producers in other Central Asian states, such as Kazakhstan. Therefore, for Uzbekistan, CA has become an area of not only cultural exchange but also extended economic activity. Importantly, Uzbekistan refuses to prioritize any particular country in CA (for instance, Kazakhstan), suggesting that it wants to build equally important relations with all Central Asian states.[19]

Within this structure, Uzbekistan still needs to develop and diversify its industrial base to modernize its manufacturing capacity and generate varied types of goods of sufficient quality to export to its neighbors. In this regard, as a third principle of Uzbek foreign policy, Mirziyoyev announced that China (along with Russia) is expected to become a non-regional partner of the highest importance for Uzbekistan. This announcement was based on previous successful engagements with China (for example, the China-led Shanghai Cooperation Organization and its "Shanghai spirit") as well as on careful calculations of the needs of the Uzbek economy, outlined in the strategy for Uzbek development explained below. Interestingly, Mirziyoyev further suggested Japan and South Korea as important partners in economic cooperation, and as sources of investments and technologies for the Uzbek economy, presumably targeting the areas where Chinese technology is not considered sufficiently developed. Currently, China is the largest export destination for Uzbekistan (Table 6.1).

Uzbek foreign behavior demonstrates that its motivations in respect to both China and CA are socially constructed and contingent on the history and legacy of previous interactions. Rationalists treat the interests of both Uzbekistan and larger powers such as China as stable (in line with domination and survival rhetoric), and predetermined, but this chapter aims to demonstrate that national interests change according to both the interaction and the environment. Therefore, both China and Uzbekistan redefine their identities (that is, become who they are) within their interactions, and their interests are defined by these interactions.

Accordingly, at the current stage of its development, Uzbekistan prioritizes CA as an area of vital importance. At the same time, the discourse of Uzbekistan siding exclusively with Russia, China or the US in the post-Karimov era was not empirically demonstrated in the first year of Mirziyoyev's presidency. From the perspective of Uzbekistan, many of the projects promoted by the EU, the US, Russia, China, South Korea and Japan are mutually compatible. In this context, China's role has shifted from being a partner in providing security and fighting terrorism to being the main source of technology and financing for the diversification of Uzbekistan's economy.

Table 6.1 Top 20 countries which have the highest trade volumes with Uzbekistan (US$ millions)

Countries	Volume of trade	Exports	Imports	Percentage of total trade	Rate of growth (%)
China	576.1	252.0	324.1	17.2	168.1
Russia	429.7	170.1	259.6	12.8	99.0
Kazakhstan	250.3	78.4	171.9	7.5	133.8
Korea	246.1	5.5	240.7	7.3	2.6
Turkey	154.6	78.6	76.0	4.6	165.6
Belarus	45.3	2.5	42.9	1.4	124.0
Turkmenistan	42.7	5.0	37.7	1.3	3.3
Germany	39.7	2.5	37.2	1.2	91.0
Lithuania	37.6	1.8	35.9	1.1	169.5
Japan	35.9	1.2	34.7	1.1	6.2
India	32.3	0.9	31.4	1.0	161.7
Ukraine	30.5	6.8	23.7	0.9	100.6
France	30.0	24.6	5.5	0.9	4.5
Latvia	30.0	2.5	27.5	0.9	85.7
Afghanistan	29.6	29.4	0.2	0.9	51.0
Tajikistan	27.2	17.2	10.0	0.8	140.8
Brazil	26.1	0.0	26.1	0.8	23.8
Kyrgyzstan	26.1	20.8	5.3	0.8	131.7
US	24.2	1.1	23.1	0.7	2.5
Iran	21.5	9.6	9.6	0.6	141.2

Source: The State Committee of the Republic of Uzbekistan on Statistics, https://stat.uz/ru/press-tsentr/novosti-komiteta/5428-vneshnetorgovyj-oborot-respubliki-uzbekistan-3, last accessed February 21, 2019.

In its interactions with foreign partners, Tashkent distinguishes long-term and short-term priorities. The long-term priorities have not radically changed. They are mainly aimed at reducing imports and increasing export-oriented manufacturing.[20] The areas prioritized by the aforementioned development strategy are rooted in an analysis of the structure of Uzbekistan's GDP. Per the latest economic indicators, the share of industry in GDP is only one-third (33.1 percent), with the remaining shares attributed to services (almost half) and agriculture, forestry, and fishery (one-fifth).[21]

Therefore, the development strategy of 2017–2021 aims to expand industry and diversify Uzbekistan's economy. Private-sector participation in the production of Uzbekistan's GDP is also very low, especially in industry, forcing the government to take the lead in seeking foreign partners.

Therefore, policy supporting the development of export-oriented, small and medium-sized private industrial production plants was announced a couple of decades ago and has been inherited by the current government. However, the short-term priorities are defined through various programs that have a timeframe for implementation. Once a specific program is completed, another defines the new short-term priorities of the country.

It would thus be incorrect to assume that Uzbekistan is prepared to accept any agenda set by its counterparts such as China or Russia, as long as this agenda implies investments. As a guideline for a cooperation agenda, the new president announced the Strategy of Actions for Further Development of the Republic of Uzbekistan for 2017–2021, which outlined five main areas to be prioritized:

- Improvements to the state public administration system.
- Improvements to the judicial system.
- Liberalization of the economy.
- Development of the social sphere.
- Facilitation of security, promotion of inter-ethnic and inter-religious accord, and implementation of a balanced, beneficial and constructive foreign policy.

The wide spectrum of reforms required has led Uzbekistan to adopt a step-by-step approach, beginning with improving the system of state operations (streamlining bureaucratic procedures for more efficient and corruption-free operations) and improving the judicial system. In respect to relations with China, "liberalization of the economy" and "implementation of a balanced and beneficial foreign policy" are of special importance. Economic investments that entail technology transfer to establish an export-oriented manufacturing sector in Uzbekistan directly relate to cooperation schemes planned with China. Such reforms are expected to prepare economic conditions in which Chinese (and other) investors will be interested in building plants in Uzbekistan. The liberalization of the currency, which implies elimination of the "black market" rate, unification of governmental and market exchange rates and guarantees by the government to convert local currency into a foreign one without restrictions, is one of the crucial steps in this direction, as it allows investors to move their capital in and out of the country freely.

The last point, regarding foreign policy, includes not only establishing priorities for countries but also drafting action plans and road maps for cooperation with each country considered strategically important. Actions plans for Kazakhstan, Russia, Turkey and Kyrgyzstan have been signed during Mirziyoyev's visits to these countries and are now being implemented. The action plan and its road map for relations with China was developed and signed during Mirziyoyev's visit in May 2017.

Infrastructure and resource development

The largest economic infrastructure–related agreements were those focusing on the joint production of synthetic fuel (US$3.7 billion), investment in Uzbekistan's oil industry (US$2.6 billion), prospective new projects (US$2 billion), cooperation in the construction of energy generation plants (US$679 million), railroad infrastructure development (US$520 million), and Tashkent to Osh (Kyrgyzstan) road construction (US$220 million). In terms of establishing manufacturing lines in Uzbekistan, agreements were reached on establishing production facilities for cement (US$153 million in Andijan and US$100 million in Tashkent), textiles (US$200 million), electric appliances (US$139 million), metals (US$115 million) and glass (US$83 million).

One of the most significant infrastructure agreements, signed in May 2017, was between Uzbekistan and the PRC on facilitating smooth international road transportation between the two countries, which involves the simplification of procedures and creation of an environment to increase the transportation of goods using land roads (inclusive of rail and motor roads). This agreement followed all the internal procedures in Uzbekistan and has been confirmed by decree of the president.[22]

As partly explained in Chapter 4, the project, which aims to connect China with Uzbekistan through the territory of the Kyrgyz Republic, has also been addressed in relevant signed agreements. It would connect the Uzbek city of Andijan and the Chinese city of Kashgar, via Osh and Irkeshtam in Kyrgyzstan by both rail and motor road. This is the shortest route, and both countries are interested in its construction.[23] While China has for years been interested in a number of transport corridors to markets in Europe through CA's transport networks, these particular rail and motor roads are of special interest and importance to Uzbekistan. They would allow Uzbekistan to shorten the distance to transport its goods into China by avoiding Kazakh railroads, which take longer and cost more. The new rail and motor roads would let Uzbekistan ship its goods directly, using the shortest route to China through Kyrgyzstan.

In 2012, Kyrgyzstan drafted its own railroad project. It would cover more areas of Kyrgyzstan and would be 380km longer than the current total. For Kyrgyzstan, it would be a chance to develop its own railroad system and connect remote areas that the rail system currently bypasses. However, for both

Uzbekistan and China, this would imply a longer transportation time for their cargo and much higher costs for the project in general. The plans suggested by Kyrgyzstan appear to be difficult for China and Uzbekistan to accept.[24]

Both Chinese and Uzbek officials realize that there is certain degree of caution (especially on the Kyrgyz side) with respect to Chinese infrastructure projects, so they require public engagement and awareness. To facilitate public acceptance of this project, the Chinese and Uzbek governments agreed to organize an auto rally along the route of the future railroad, which serves several important goals. First, it aims to promote to the public transportation infrastructure development projects between Chinese and Central Asian counterparts. Second, it is a practical test of the preparedness of areas where railroad construction is planned, to determine any issues in the terrain and to detail the infrastructure-related facilities that must be constructed in this area. In addition to the railroad, the Chinese Railway Tunnel Group, which built the 19-kilometer-long Kamchik Tunnel in Uzbekistan, has committed to construction of a vehicle motorway under the tunnel, naming the project Kamchik 2.[25]

Uzbekistan's exports of mineral and natural resources to China constitute a considerable share of the trade between the countries. According to other agreements concluded in May 2017 during Mirziyoyev's visit to China, contracts identified natural gas (six billion cubic meters, worth US$734 million), uranium (US$30 million), textiles (US$200 million), leather (US$21.3 million), and agricultural products (US$1.6 million) as products to be exported to China by the end of 2017. Plans have also been articulated for exports of natural gas to reach US$2.4 billion for the years 2018 to 2020.[26]

There were also discussions of new pipelines to connect natural gas–endowed Turkmenistan, Uzbekistan and Kazakhstan to Chinese consumers to ensure a stable supply. However, no construction plans or financial commitments have yet been achieved due to uncertainties regarding the economic sustainability of the pipelines' operations.

Chinese companies, such as Shengli Oil and Freet Petroleum, represent strategic partners for the government and related corporations in the provision of pipes and other extraction equipment. According to sources aware of the negotiations between the government and Chinese corporations, at least three contracts for the supply of such equipment were signed during Mirziyoyev's visit to Beijing in 2017.

China National Petroleum Corporation secured a co-financing contract from the Bank of China for a project for drilling at the gas condensate field in Bukhara, Uzbekistan by establishing JV New Silk Road Oil and Gas, which was set up by Uzbekneftegaz and China National Oil and Gas Exploration and Development Corporation (a subsidiary of China National Petroleum Corporation).[27] According to the license granted to the joint venture, it plans to develop the existing wells and drill another 16, with annual production to reach 1 billion cubic meters of natural gas and 6,500 tons of condensate.[28]

In terms of the generation of new industries, Uzbekistan concluded an agreement between Uzbekneftegaz and the Chinese Development Bank (worth US$3.7

billion, of which US$1.2 billion is to be financed by China) to finance the establishment of a plant to produce synthetic fuel at Uzbekistan's largest gas refinery complex, Shurtan.[29] The plant would process 3.6 billion cubic meters of natural gas into 743,000 tons of synthetic fuel, 311,000 tons of aviation fuel, 431,000 tons of naphtha fuel, and 20,900 tons of liquefied gas.[30] Technological support for the plant is to be provided by South Korea's Hyundai Engineering & Construction under a license provided by South African Sasol. The technology for turning natural gas into liquefied gas would come from Dutch Haldor Topsoe.

A US$3 billion agreement between the Chinese Ministry of Commerce and Uzbekgidro stipulates the modernization of approximately 300 water pump stations to improve the efficiency of Uzbekistan's hydroelectricity sector.[31] Uzbekistan developed and approved the strategy for hydro-energy development in November 2015 and aims to invest US$889.4 million in the sector between 2016 and 2020. The modernization of 15 hydroelectric stations is planned, to allow the generation of 5.25 billion kilowatts of energy. This governmental program was essential in defining this area as a high priority in relations with China.

Modernization of energy generation in Uzbekistan has also been prioritized in negotiations. In particular, the China Railway Tunnel Group (CRTG) and China Coal Technology & Engineering Group began the modernization of a plant to achieve extraction of 900,000 tons of coal per year, with investment of US$94.5 million.[32] Non-traditional sustainable sources such as biomass generation have also been the subject of agreements. Uzbekneftegaz, AKB Agrobank, and China's Poly International Holding signed a memorandum of cooperation to set up the production of modern biogas plants worth US$10 million and to assist in the modernization of eight domestic enterprises, including the JSC Oil and Gas and Chemical Engineering Plant, in line with the governmental Program of Measures to Increase Biogas Plants in Uzbekistan for 2017–2019.[33] Last but not least, solid waste processing infrastructure is being constructed to improve livelihoods and facilitate better waste utilization in old and newly constructed quarters in Uzbekistan.[34]

Manufacturing and export industries

As described above, the new administration in Uzbekistan has defined certain areas where China's economic presence and technology are desired and advantageous. In line with the goals of the development strategy, the first such area is the establishment of small and medium-sized manufacturing plants to not only supplement imported products but also, importantly, produce goods that can compete in Central Asian markets.

In particular, Uzbekistan intends to establish long-term cooperation with the Chinese Aerospace Science and Industry Corporation (CASIC) for the supply of scanning equipment for border control services, customs, airports and rail infrastructure, the introduction of urban security systems ("smart city"), the protection of large facilities and the border, the development and introduction of an industrial Internet, joint production of a wide range of pharmaceuticals in Uzbekistan, and the participation of CASIC in the development of infrastructure for the Central Military Clinical Hospital of the Ministry of Defense. This latter project includes creating a turnkey multidisciplinary medical and diagnostic building and equipping it with modern medical equipment, as well as producing oil and gas equipment.[35]

To respond to the increasing demand for construction materials, the Uzbek government intends to facilitate the development of a cement plant (with Zhejiang Shangfeng Building Materials, at a cost of US$203.9 million)[36] and glass production (with MingYuan Silu, at a cost of US$110 million).[37] Another joint venture has been established in Gulistan City, focusing on the production of elevators. In the Soviet era, Uzbekistan relied heavily on elevators produced in other republics, namely Azerbaijan and Russia. With the collapse of the Soviet Union, the replacement of installed elevators required importing them in great numbers. To fill this gap and respond to the increasing need for elevators in the booming construction business in Uzbekistan, the government sought China's assistance in facilitating the production of elevators in Uzbekistan. As a result, a joint venture for the production of elevators (Modern Lift Systems) has been established in the Syrdarya Region of Uzbekistan, funded by Chinese investors and using Chinese technology. It produces 300 elevators per year, plus 200 escalators and travelators (moving walkways). Although the joint venture produces elevators for internal consumption, approximately 30 percent of its products are to be exported to other regional states.[38]

In terms of memoranda and protocols, some of these have already materialized in production facilities, such as the one for the production of soft and hard toys in Tashkent, based on the Soviet-era toy factory Tashkentigrushka.[39] As mentioned, Uzbekistan is the most populous country in CA, and 60 percent of its 32 million people (2017) are under 25. The population has the highest growth rate in CA, which creates a big market for toys and child-related products. In the Soviet era, Uzbekistan hosted one of the biggest toy factories. But with the collapse of the Soviet Union and aging technology, the plant became unable to meet the needs of the market. The quality of toys imported from other countries, including China, was not at the level expected by consumers, motivating the government to seek a solution to this issue. Imports of toys reached US$2.1 million in 2016, but fell 29 percent in 2017 due to increasing local production. As mentioned, 94 percent (US$2 million) of the toys imported into Uzbekistan are from China; Russia's share is only 2.1 percent (US$43,000), with the remaining 0.8 percent (US$16,000) from Lithuania.[40] The new factory (jointly established with Shandong Sanhe Toy) cost an estimated US$23 million, makes 700 kinds of plastic, soft, electronic and mechanical toys, and provides 950 jobs in Tashkent.[41]

The majority of these projects attempt to establish production- and infrastructure-related facilities to enable Uzbekistan to produce goods not only for its large (but still limited) internal consumption but also, importantly, for export. The toy factory aims to produce seven million individual toys annually, of which 80 percent are meant for export to other countries of CA, Russia, Afghanistan and beyond. Similar plants are also planned in conjunction with other Chinese producers (such as Zhejiang Jiyou) in the Jizzakh free economic area.[42]

Similar protocols for intentions to establish production plants for ceramics with Peng Yu Special Ceramics and Jingdezhen Porcelain, and a porcelain production plant with Ru Hong, have been signed.[43] Although the documents signed with representatives of these companies were protocols of intention without firm commitments, in June and July 2017 the representatives visited possible sites for plant construction and discussed conditions with various ministries, such as the Ministry of Foreign Trade, the Foreign Investments Agency and the State Committee for Competition Controls.

Among the agreements signed between the two countries, Sun Chapter Industry and China National Complete Plant Import & Export signed a protocol of intentions with the government to create a cluster for the production of high-quality paper in the Angren free economic zone.[44] Uzbekistan currently does not have such facilities and needs to import much of the high-quality paper used in office paperwork, as well as for wrapping and shipping the goods produced in various plants.[45] Delegations from these companies visited Uzbekistan in September 2017 to evaluate the needs and coordinate the supplies of equipment with local counterparts.

In terms of the export of Uzbek-made products, the agreements signed during the May 2017 visit of President Mirziyoyev included textile exports (US$300 million), leather (US$60 million), and agricultural products (US$1.6 million) in 2017–2018. These contracts are in active implementation, and the majority are being implemented. Also, cotton processing and textile mills have been planned, with Chinese participation, in Qashqadaryo for 2017–2019. Seven textile mills are under construction in Qashqadaryo, and their overall cotton processing capacity will account for 10 percent of the annual cotton output of the Qashqadaryo Region.[46] The Litai project aims to create 500 jobs, using textile machinery from the Saurer Group, and to produce 22,000 tons of high-level cotton yarn annually, 80 percent for export.[47] Wenzhou Jinsheng Trading announced that it will initiate seven investment projects in the Jizzakh economic zone, investing US$40 million in reprocessing local resources and producing leather goods and metal products, one-third to be exported from Uzbekistan.[48]

Conclusions

A few conclusions can be drawn from the analysis of cooperation between Uzbekistan and China in this chapter. Uzbekistan's relations with China in the year after the death of President Karimov and the opening up of Uzbekistan to the international community demonstrate that the master narrative overestimates Chinese domination in both agenda setting and project implementation in areas of economic cooperation. Although both China and Uzbekistan are attempting to shape their new foreign policies in a changing international environment, most of the projects detailed in this chapter focus on areas that are defined as highly important for either Uzbekistan or China. It does not appear that China is using its economic power to pressure smaller Uzbekistan. There are several factors that allow Uzbekistan to maintain its relations with China without sacrificing its own interests. First, in case one or more projects collapse, Uzbekistan has been diversifying its strategic partners and building ties with alternative countries. This provides a diverse selection of partners for particular developmental goals. It also keeps some pressure on China not to dominate the cooperation agenda if it sees value in engaging Uzbekistan.

Second, Sino-Uzbek relations demonstrate that neither China's nor Uzbekistan's foreign policy behavior has a static structure or goal. Analysts and observers tend to emphasize the dominating tendencies of Chinese economic expansion (by emphasizing energy resources) or its agenda for negotiations. The projects described in this chapter demonstrate that a majority of the cooperative initiatives are products of social construction between the main actors and their negotiations. The funded and facilitated projects in Uzbekistan do not represent a "one-way street" of Chinese expansionist policy.

Third, the outcomes of this cooperation are greatly influenced by the respective government bureaucracies and their facilitation of inter-state cooperation. In countries where the governments play similar proactive roles in defining economic policies, the role of bureaucracies is instrumental in ensuring productive cooperation. In Uzbekistan's policy in respect to China, its strategy for development and its overall goals of diversifying the economy and orienting industry toward exports serve as markers for the Uzbek bureaucracy when deciding priorities. Those guidelines and bureaucratic procedures of agenda setting assist in streamlining the agenda of cooperation with China in a desirable direction, not only for Chinese interests but also for the cooperating partner, in this case Uzbekistan.

Notes

1 See e.g., "Uzbekistan I Kitaj gotovy prodvigat stroitel'stvo novogo Evrazijskogo kontinental'nogo mosta" [Uzbekistan and China are ready to promote construction of new Eurasian land bridge], Interview with Ambassador of China to Uzbekistan Sun Lijie, *Podrobno*, June 2, 2017; Oybek Madiyev, "Why Have China and Russia Become Uzbekistan's Biggest Energy Partners? Exploring the Role of Exogenous and Endogenous Factors", *Cambridge Journal of Eurasian Studies* 1, March 15, 2017, https://doi.org/10.22261/QYJ7IT.

2 For a narrative of Chinese interests in CA, see Andrew Scobell, Ely Ratner and Michael Beckley, *China's Strategy toward South and Central Asia: An Empty Fortress* (Santa Monica, CA: RAND Project Air Force, 2014): 27–48, www.rand.org/pubs/research_reports/RR525.readonline.html, accessed September 23, 2017.

3 On the complementarity of Chinese Belt and Road and Kazakh domestic Nurly Zhol policy, see e.g., Nargis Kassenova, "China's Silk Road and Kazakhstan's Bright Path: Linking Dreams of Prosperity", *Ponars Eurasia*, 2017, www.ponarseurasia.org/article/china's-silk-road-and-kazakhstan's-bright-path-linking-dreams-prosperity, accessed October 2, 2017.

4 On Chinese domestic agenda setting in foreign policy, see Linda Jakobson and Ryan Manuel, "How Are Foreign Policy Decisions Made in China?", *Asia & the Pacific Policy Studies* 3, no. 1 (2016): 101–110, doi:10.1002/app. 5.121.

5 On increasing Chinese global assertiveness, see Nien-Chung Chang Lao, "The Sources of China's Assertiveness: The System, Domestic Politics or Leadership Preferences?", *International Affairs* 92 (2016): 817–833, doi:10.1111/1468–2346.12655.

6 National Development and Reform Commission, Ministry of Foreign Affairs and Ministry of Commerce of People's Republic of China, "Vision and Actions on Jointly Building Silk Road Economic Belt and 21st-century Maritime Silk Road", March 30, 2015; Office of the Leading Group for the Belt and Road Initiative, *Building the Belt and Road: Concept, Practice and China's Contribution* (Beijing: Foreign Languages Press, May 2017), https://eng.yidaiyilu.gov.cn/wcm.files/upload/CMSydylyw/201705/201705110537027.pdf, accessed March 12, 2018; Timur Dadabaev, " 'Silk Road' as Foreign Policy Discourse: The Construction of Chinese, Japanese and Korean Engagement Strategies in Central Asia", *Journal of Eurasian Studies* 9, no. 1 (2018): 30–41. doi:10.1016/j.euras.2017.12.003.

7 Timur Dadabaev, "The Constructivist Logic of Uzbekistan's Foreign Policy in the Karimov Era and Beyond", Uzbekistan Forum and Virtual Special Issue, *Central Asian Survey* (September 30, 2016): 1–4, www.tandf.co.uk//journals/pdf/8_Dadabaev_final%20220916.pdf, last accessed June 9, 2018); Also see Timur Dadabaev, "Uzbekistan as Central Asian Game Changer? Uzbekistan's Foreign Policy Construction in the Post-Karimov era", *Asian Journal of Comparative Politics* 4, no. 2 (2018, print forthcoming, 2019): 162–175, doi:10.1177/2057891118775289.

8 Samuel Ramani, "Are Uzbekistan's Ties with China Headed for a Change?", *Radio Free Europe/Radio Liberty*, September 12, 2016, www.rferl.org/a/qishloq-ovozi-uzbekistan-china-relations-change-karimov-death/27982369.html, accessed September 23, 2017.

9 Sarah Lain, "Russia and China: Cooperation and Competition in Central Asia", in *Chinese Foreign Policy under Xi*, ed. Tiang Boon Hoo (London: Routledge, 2017), pp. 74–95; Timur Dadabaev, "Engagement and Contestation: The Entangled Imagery of the Silk Road", *Cambridge Journal of Eurasian Studies* 2018, no. 2, doi:10.22261/CJES.Q4GIV6.

10 Shavkat Mirziyoyev, "Effektivnaya Venshnyaya Politika: Vazhneishee Uslovie Uspeshnoy Realizatsii Vneshnei Politiki" [Effective foreign policy: The most important condition of the successful foreign policy], January 11, 2018, http://press.natlib.uz/ru/edition/download?id=4558, last accessed June 9, 2018.

11 See e.g., Jing-Dong Yuan, "China's Role in Establishing and Building the Shanghai Cooperation Organization (SCO)", *Journal of Contemporary China* 19, no. 67 (2010): 855–869; Timur Dadabaev, "Shanghai Cooperation Organization (SCO) Regional Identity Formation from the Perspective of Central Asian States", *Journal of Contemporary China* 23, no. 85 (2014): 102–118, doi:10.1080/1067056 4.2013.809982.

12 In particular, the Ministry of Agriculture and Water Management of Uzbekistan is cooperating with Japan on increasing the efficiency of water consumption at the consumer level. The ministry also announced in 2017 that Uzbekistan is interested in importing technologies using hydroponic growing methods for agricultural products from Japan, South Korea, and Iran ("RUz Budet Vyraschivat Frukty Kruglyi God", *Sputnik*, September 6, 2017, http://ru.sputniknews-uz.com/economy/20170906/6236626/uzbekistan-selhozkulturi.html, accessed September 23, 2017). Hydroponic growing uses mineral nutrient solutions to feed the plants in water, without soil. This technology is important for Uzbekistan, given the water deficiencies and soil corrosion in certain areas of the country. It could also allow Uzbekistan to produce agricultural products throughout the year. Finally, India and Pakistan have been given roles as trading and humanitarian exchange partners.

13 Bernardo Teles Fazendeiro, *Uzbekistan's Foreign Policy: The Struggle for Recognition and Self-Reliance under Karimov* (Oxon: Routledge, 2017).

14 Perceived threats from Turkey included support to the Anti-Karimov opposition in exile and influence of Islamic radicals from Turkey on internal politics of Uzbekistan.

15 Timur Dadabaev, *Towards Central Asian Regional Integration: A Scheme for Transitional States* (Tokyo: Akashi, 2004).

16 "Andijan events" refers to the May 13, 2005 clash between the Uzbek Interior Ministry and National Security Service and a crowd of protesters in Andijan, whom government claimed were an Islamic radical group called Akromiya, during which a great number of civilian casualties were reported.

17 Asatryan, Zabrodin, and Sozaev-Gur'ev, "V Moskve Zhdut Uzbekskogo Lidera i Raschityvayut na Konstruktivnyi Dialog" [Moscow expects Uzbek leader and hopes for constructive dialogue], *Izvestia*, December 7, 2016, http://iz.ru/news/649838, accessed April 2, 2017.

18 See also Falyahov, "Uzbekistan ne Speshit v Evrazijskij Soyuz" [Uzbekistan does not rush into Eurasian Union], *Gazeta*, April 5, 2017, www.gazeta.ru/business/2017/04/05/10612535.shtml, accessed September 23, 2017.

19 "Glava MID Uzbekistana otvergnul ideyu sozdaniya 'regional'noi osi' Astana-Tashkent" [The Head of MFA of Uzbekistan rebuffed the idea of creating 'regional axis' of Astana-Tashkent], *Podrobno*, September 29, 2017, http://podrobno.uz/cat/politic/glava-mid-uzbekistana-otvergnul-ideyu-sozdaniya-regionalnoy-osi-astana-tashkent/, accessed September 29, 2017.

20 On the logic of Uzbek self-reliance, see Bernardo Teles Fazendeiro, "Uzbekistan's 'Spirit' of Self-Reliance and the Logic of Appropriateness: TAPOich and Interaction with Russia", *Central Asian Survey* 34, no. 4 (2015): 484–498, doi:10.1080/02634937. 2015.1114780.

21 State Statistics Committee, "Statesticheskoe Obozrenie Uzbekistana" [Statistic review of Uzbekistan], Tashkent, 2017: 4.

22 Decree of the President of Uzbekistan on Reconfirming the "Agreement between Republic of Uzbekistan and People's Republic of China on Facilitating Smooth International Road Transportation", PP-3143, July 20, 2017, in Russian at http://nrm.uz/contentf?doc=508717_postanovlenie_prezidenta_respubliki_uzbekistan_ot_20_07_2017_g_n_pp-3143_ob_utverjdenii_mejdunarodnogo_dogovora&products=1_vse_zakonodatelstvo_uzbekistana, accessed September 23, 2017.

23 For details, see A Titova, "Uzbekistan hochet postroit Andijan-Osh-Irkishtam-Kashgar" [Uzbekistan wants to build Andijan-Osh-Irkishtam-Kashgar], *Kloop*, September 9, 2017, https://kloop.kg/blog/2017/09/09/uzbekistan-hochet-postroit-trassu-andizhan-osh-irkeshtam-kashgar/, accessed September 23, 2017.

24 See e.g., Bruce Pannier, "No Stops in Kyrgyzstan for China-Uzbekistan Railway Line", *Radio Free Europe/Radio Liberty*, September 3, 2017, www.rferl.org/a/qishloq-ovozi-kyrgyzstan-uzbekistan-china-railway/28713485.html, accessed September 23, 2017.

25 "China to Help Build Second Tunnel at Kamchik Pass", *Tashkent Times*, May 16, 2017, www.tashkenttimes.uz/economy/930-china-to-help-build-second-tunnel-at-kamchik-pass, accessed September 23, 2017.

26 "Uzbekistan planiruet k 2021 godu narastit' export gaza v Kitaj do 10 mlrd" [Uzbekistan plans to increase exports of gas to China to 10 billion cubic meters until 2021], *Interfax News Agency*, June 15, 2017, http://interfax.az/view/705802, accessed September 23, 2017; "Uzbekistan nameren narastit postavki gaza v Kitai" [Uzbekistan aims to increase the volume of its gas exports to China], *Aziay Plus*, June 15, 2017, https://news.tj/ru/news/tajikistan/economic/20170615/uzbekistan-nameren-narastit-postavki-gaza-v-kitai, accessed September 23, 2017.

27 "Uzbekistan–China JV New Silk Road Oil and Gas Commences Drilling in Bukhara", *Tashkent Times*, June 15, 2017, www.tashkenttimes.uz/economy/1016-uzbekistan-china-jv-new-silk-road-oil-and-gas-commences-drilling-in-bukhara, accessed September 23, 2017.

28 "Uzbekistan–China JV New Silk Road Oil and Gas".

29 For details, see George Voloshin, "Central Asia Ready to Follow China's Lead Despite Russian Ties", *Eurasia Daily Monitor* 14, no. 71 (2017), https://jamestown.org/program/

central-asia-ready-follow-chinas-lead-despite-russian-ties/, accessed September 23, 2017; "Uzbekistan: President's China Trip Yields Giant Rewards", *Eurasianet*, May 18, 2017, https://eurasianet.org/s/uzbekistan-presidents-china-trip-yields-giant-rewards, accessed June 9, 2018.

30 "Uzbekistan i Kitai podpisali soglazhenij na summu bolee 20 mlrd" [Uzbekistan and China signed agreements for more than 20 billion], *Review.uz*, May 14, 2017, www.review.uz/novosti-main/item/11217-uzbekistan-i-kitaj-podpisali-soglashenij-na-summu-bolee-20-mlrd, accessed September 23, 2017.

31 "Uzbekistan I Kitai podpisali soglazhenij na summu bolee 20 mlrd".

32 "Startovala modernizatsiya predpriyatiya 'Shargunkomir'" [The modernization of 'Shargunkomir' has started], *Gazeta*, September 29, 2017, www.gazeta.uz/ru/2017/09/07/coal/, accessed September 23, 2017.

33 "Program for Increased Use of Biogas in Farms Adopted in Uzbekistan," *Tashkent Times*, June 19, 2017, www.tashkenttimes.uz/economy/1060-program-for-increased-use-of-biogas-in-farms-adopted-in-uzbekistan, accessed September 23, 2017.

34 Beston Machinery Company, "Beston Municipal Solid Waste Plant in Uzbekistan", undated, http://mswrecyclingplant.com/beston-municipal-solid-waste-plant-uzbekistan/, accessed September 23, 2017.

35 For details, see "Ministry of Foreign Trade Discuss Cooperation with 'CASIC' Corporation", *Uzreport*, June 16, 2017, http://news.uzreport.uz/news_4_e_153308.html, accessed September 23, 2017.

36 "New Cement Plant for US$204 Million to be Built in Andijan Region", *Uzdaily*, July 21, 2017, https://uzdaily.com/PYLnM/articles-id-40176.htm, accessed July 21, 2017.

37 "Glass Production Plant to be Considered in Uzbekistan", *Uzdaily*, September 21, 2017, www.uzdaily.com/articles-id-37934.htm, accessed September 23, 2017.

38 "Proizvodtsvo liftov zapuscheno v Uzbekistane" [Production of Elevators launched in Uzbekistan], *Gazeta*, September 29, 2017, www.gazeta.uz/ru/2017/09/29/lifts/, accessed September 23, 2017.

39 "O merakh po organizatsii sovremennogo proizvodstva detskih prinadlezhnostei i igrushek v gorode Tashkente" [Decree of the President of the Republic of Uzbekistan on measures for organization of production of children's accessories and toys in the city of Tashkent], PP-3115, July 6, 2017.

40 "Uzbekistan and China Will Create Production of Joint Venture on Production of Toys", *Gazeta*, August 2, 2017, www.gazeta.uz/ru/2017/08/02/toys/, accessed August 4, 2017.

41 For details, see "Klaster po vypusku igrushek stoimost'yu 23 mln pyavitsya v Tashkente" [Cluster producing toys costing 23 mln will appear in Tashkent], *Gazeta.uz*, September 13, 2017, www.gazeta.uz/ru/2017/09/13/cluster/, accessed September 14, 2017.

42 "Zhejiang Jiyou Industrial Co. Ltd. May Create Capacities for Toy Production in Uzbekistan", *Uzdaily.uz*, July 3, 2017, www.uzdaily.com/articles-id-39970.htm, accessed September 23, 2017.

43 "FEZ 'Angren' May Become the Largest Producer of Porcelain in the Country", *UzbekistanToday*, July 3, 2017, http://ut.uz/en/business/fez-angren-may-become-the-largest-producer-of-porcelain-in-the-country/, accessed September 23, 2017.

44 "COMPLANT planiruet organizovat' proizvodstvo bumagi v Uzbekistan" [COMPLANT plans to organize production of paper in Uzbekistan], *Obschestvo Druzhby Kitaj-Uzbekistan*, September 18, 2017, http://china-uz-friendship.com/?p=12945, accessed September 23, 2017.

45 "V Uzbekistane budet sozdano SP po proizvodstvu bumagi" [Uzbekistan will create joint venture on production of paper], *Uzreport*, September 18, 2017, https://uzreport.news/economy/v-uzbekistane-budet-sozdano-sp-po-proizvodstvu-bumajnoy-produktsii, accessed September 23, 2017.

46 "Uzbekistan President Visits Litai Silk Road Park", *Jinsheng*, March 4, 2017, www. jinshengroup.com/en/stations/532661b60b/index.php/5327ed9b0a?id=161, accessed September 23, 2017.
47 "Uzbekistan President Visits Litai Silk Road Park".
48 "Kitaj pomozhet realizovat v SEZ 'Jizzakh' proekty stoimost'yu $40 millionov" [China will help implement projects worth 40 million in Jizzakh SEZ], *Sputnik*, August 11, 2017, http://ru.sputniknews-uz.com/economy/20170811/6014975/kitai-sez-djizak-proekti.html, accessed September 23, 2017.

7   Revisiting Japan's Silk Road master-narratives

In recent years, the term "Silk Road" has been appropriated by the Chinese Belt and Road Initiative. Consequently, any reference to the Silk Road now has connotations of Chinese penetration and engagement of the Central Asia (CA) region and beyond. However, one of the first countries in East Asia that applied the notion of the Silk Road to its diplomatic initiatives in CA was Japan. The Japanese usage of "Silk Road" was then repeated by the US, which proposed the Silk Road Strategy Act of 1999 to expand the US presence in this region and to restrain Russian dominance. South Korea subsequently also launched a number of similar strategies through 2009–2013 under the "Silk Road" umbrella, to connect South Korea through Russia, China and the CA railroad networks to energy and other resources in Eurasia. India, Iran, Turkey and other countries have also undertaken various initiatives, all of which demonstrates the international environment and contested nature of the CA engagement in which Japan operates.

Japan has since solidified its presence in CA and contributed significantly to regional development through its ODA assistance, which, for various reasons, remains in the shadow of Chinese infrastructure construction projects (Belt and Road Initiative and construction of "land bridges" from China to Europe through the CA region, as well as exports of CA natural gas and oil from Kazakhstan, Uzbekistan and Turkmenistan) or Russian initiatives (the Eurasian Economic Community, which implies creation of a customs and economic union).

With the focus on Japanese foreign policy in CA, this chapter aims to clarify the following issues. First, it attempts to explain the process of the construction of Japanese diplomatic initiatives in the CA region in historical perspective. Second, this chapter highlights the strengths and weaknesses of the road maps between Uzbekistan and Japan worked out by the joint intergovernmental committee, as well as the areas of Japanese interest that these road maps detail. And third, the chapter sheds light on the areas in need of attention from the Japanese policy makers and experts in their efforts to make the Japanese engagement in this region more efficient.

Evolution of the Japanese Silk Road narrative

As has been described in several chapters of this volume, the Japanese involvement in the CA region and in Uzbekistan has been diverse. It has become one of the most significant assistance providers and supporters of state-building initiatives. It has also provided much needed technical assistance and loans to regional states. In addition, Japan has initiated the process of regional dialogue between regional states and with Japan itself.

At the same time, there is a significant expectation on the side of CA governments and the public in respect to Japan. For CA governments, Japanese involvement in CA represents an attempt to balance Russian and Chinese engagements, while having access to the technologies and knowledge much needed to upgrade their industrial base. The public perception of Japanese influence on CA states is generally positive. According to the AsiaBarometer poll conducted by the University of Tokyo in the autumn of 2005, the highest ratings for good and rather good influence of Japan on their country registered in Kazakhstan (10.4 percent and 30.3 percent respectively) and Uzbekistan (15.9 percent and 36.3 percent respectively). Similarly, a poll conducted in 2015 by the Japanese Ministry of Foreign Affairs (MOFA) prior to the autumn visit of PM Abe to CA demonstrated that sympathy toward Japan registered in 2005 had been sustained throughout the intervening 10 years, with the majority of respondents considering their countries' relations with Japan to be good (Uzbekistan 79 percent, Tajikistan 56 percent, Kyrgyzstan 52 percent and Kazakhstan 59 percent) or rather good (Uzbekistan 13 percent, Tajikistan 24 percent, Kyrgyzstan 23 percent and Kazakhstan 42 percent).[1]

Such sentiments can be attributed to the fact that CA states never had the issues related to the imperial history of Japan that are seen in the relations of Japan with East Asian countries. For them, Japan is not associated with imperialism, but rather with technological progress, proper manners and the years of ODA commitments given by Japan at the onset of the CA states' independence. Such sympathy toward Japan spills over into expectation of larger Japanese direct foreign investments and corporate participation as opposed to ODA disbursements. In addition, Japan is considered as offering an alternative to projects from China and Russia, countries potentially feared by smaller CA states because of the potential for political and economic exploitation and domination.

As described in the previous chapters, Japanese diplomacy initiatives in the last 28 years after the collapse of the Soviet Union have aimed to rediscover this latest Asian frontier for Japan and establish a Japanese presence in this region. Japanese Silk Road Diplomacy launched in 1997 and became one of the first international diplomatic initiatives appealing to the connectivity and revival of the Silk Road. This was undertaken under Prime Minister (PM) Hashimoto Ryutaro's administration. Hashimoto's understanding of this region has been informed by the Obuchi Mission[2] and Hashimoto's interactions primarily with Russia. Despite launching the Silk Road/Eurasian Diplomacy, Hashimoto never traveled to CA and the Caucasus, which partly reflects the focus of Japanese foreign policy toward the US, defined by the two countries' strategic alliance; toward China, due to Japan's economic commitments there; and toward Russia, directed at resolving territorial disputes. Such a foreign policy agenda constrained Japan's prime ministerial visits and did not leave much space for other regions, including CA.

Competitive advantages of the Japanese standing in CA

There are certain features that Japan aims to use as its competitive advantage in approaching the CA region. First, Japan's relative distance from the region is frequently interpreted as a weakness due to logistical problems associated with reaching regional resources and markets. However, the Japanese government aims to use this as a competitive advantage (when compared to other countries such as China and Russia), because this allows the Japanese government to claim "selfless" commitment to the region by suggesting that its distant geographic location prevents it from dominating and exploiting CA states. Such a claim of "altruism" in CA engagement, genuine or perceived, is part of the discursive construction of competitive advantage in respect to other big players, such as China and Russia.

Second, Japan aims to emphasize the decolonization mission of its regional institution building, exemplified by the "CA plus Japan" dialogue forum. In doing so, the Japanese government emphasizes that this scheme was designed to encourage CA states to seek intra-regional cooperation and ties with each other, while Japan would provide the technical and financial assistance needed to support such alliances. This objective of the CA plus Japan initiative is rooted in the legacy of PM Hashimoto's Eurasian (Silk Road) Diplomacy.

Third, Japan uses the duality (implied by universal and Asian features) of its identity, mentioned in Chapter 3 of this study, to forward constructive relations with CA states. One point which needs to be mentioned here is the fact that Japan is regarded in CA as being a modern society that in the past challenged the West but then became part of it without losing its traditional values, which appeals particularly to the Turkic and Muslim world that CA represents after their being first under Soviet rule and then under the shadow of potential Chinese economic dominance in recent years. In addition, Japan does not completely abandon its commitment to universal values (such as democracy, a market economy, the safeguarding of human rights and the rule of law) but also does not use it as precondition for cooperation, offering CA states an opportunity to adjust and build their domestic conditions for implementation of these universal values. To exemplify this point, one should mention that PM Koizumi was the first leader of the liberal democratic country who visited Uzbekistan in 2006 when the US and other European nations were introducing sanctions against Islam Karimov's government for excessive use of force and the eventual massacre of protesters in the city of

Andijan in May of 2005. During the visit, PM Koizumi did bring up the importance of human rights but also expressed understanding of developmental concerns of Uzbekistan and pledged unconditional developmental assistance. Thus, Uzbekistan, despite its poor human rights record has ever since been one of the top recipients of the Japanese aid which in 2016 reached 38,898 million yen constituting 73.8 percent of all the ODA assistance extended through JICA to the countries of CA and the Caucasus (see Table 7.1). In regards to regional states, Japan has ranked in the top five ODA providers for the 2010–2015 period, being the top assistance provider for Uzbekistan, and ranging between second and third top assistance provider for Kyrgyzstan and Tajikistan over these years (see Table 7.2 for overall disbursements).

A similar example of duality concerns good governance and transparency in Japan's ODA practices. Some of its companies (in particular Nihon Koutsu Gijyutsu [Japan Transportation Consultants][3]) were caught paying bribes to Uzbek (and Vietnamese and other states') officials during the process of deciding on ODA disbursements. Although Japan condemned such practices,[4] it nevertheless continued its ODA assistance without imposing punitive measures against the Uzbek government, again displaying a certain degree of understanding in respect to various problems of transition within CA. In this sense, such duality of approach represents pragmatism in the Japanese approach toward CA states and the Japanese understanding that issues of governance and transparency can only be dealt with through constructively engaging CA states over long-term engagements and not through sanctions/ punitive measures. As a channel for influencing the behavior of these states, Japan emphasizes human resource development, in line with which, by 2014, Japan had accepted 10,878 trainees from CA and the Caucasus and dispatched 2,603 experts to these states.[5]

The Japanese direct investments and commitments are less impressive, leaving considerable potential still to be fulfilled. The largest investments were made in economically larger and energy resource–rich Uzbekistan (US$900 million) and Kazakhstan (US$357 million), according to 2017 data. Japan largely exports to CA states machinery and industrial goods (to the value of 13.56 billion yen to Uzbekistan, 30 billion yen to Kazakhstan, 662 million yen to Tajikistan and 2.4 billion yen to Kyrgyzstan) while it imports from them textile yarn, fabrics and nonferrous metal (from Uzbekistan for 500 million yen), radioactive material and nonferrous metals (from Kazakhstan for 141.1 billion yen), fruits and non-metallic ware (from Tajikistan for 163 million yen and Kyrgyzstan for 153 million yen).

In terms of major actors in these interactions, the intergovernmental committee on economic cooperation between Japan and CA states is composed of the Japanese MOFA, the Japan Association for Trade with Russia & NIS (ROTOBO), JICA, the Trade Chamber and representatives of Japanese corporations (see Figure 5.7).

Table 7.1 Japanese ODA offered to the countries of Central Asia on a bilateral basis by country (US$ millions)

CA country	2001	2002	2003	2004	2005	2006	2007	2008	2009	2010	2011	2012	2013
Uzbekistan	30.92	40.16	63.22	99.75	60.02	29.60	70.29	64.53	41.92	34.08	31.26	26.25	56.49
Kazakhstan	43.93	30.13	136.27	134.34	69.68	28.19	55.39	56.63	63.38	30.56	19.79	30.89	36.99
Kyrgyz Rep.	23.15	8.12	31.23	26.69	20.95	17.22	15.69	12.49	18.06	23.50	30.99	19.98	17.87
Tajikistan	4.61	26.96	4.77	6.58	9.93	8.04	9.43	8.06	26.24	43.42	35.59	32.98	26.66
Turkmenistan	16.42	11.37	6.80	2.22	0.13	0.62	0.38	0.57	1.15	1.55	1.27	0.53	0.56

Source: Compiled from the data made available by the Ministry of Foreign Affairs of Japan, *Seifu Kaihatsu Enjyo (ODA) Kunibetsu de-tabuku 2014 (Chuou ajia/ kokasasu chiiki)* [Official Development Assistance by-country data-book 2014 (Region of Central Asia and Caucasus)]. Tokyo, Japan, www.mofa.go.jp/mofaj/gaiko/ oda/files/000072593.pdf, last accessed on July 15, 2015.

Table 7.2 JICA disbursements to CA (in millions of yen)

CA country	Total Value of JICA programs	Composition ratio %
Uzbekistan	38,898	73.8
Tajikistan	3,349	6.4
Kyrgyz Republic	2,948	5.6
Kazakhstan	155	0.3
Turkmenistan	22	0

Source: Modified by author to include only countries of CA from JICA, *East Asia and Central Asia: Towards Sustained Economic Development through Strengthening Regional Connectivity and Diversifying Industries*, JICA Activity Reports (Tokyo: JICA, 2016).

The economic cooperation road maps produced by this intergovernmental committee mostly consist of intergovernmental framework agreements and mutual understanding memoranda and agreements regarding official development assistance projects, mainly because the Japanese enterprises have not yet expressed an overwhelming commitment to involvement in projects in CA. Japanese corporations remain rather passive in CA due to concerns related both to the protection of their prospective investments and to the governance practices demonstrated in the case of Japan Transportation Consultants mentioned earlier. To put it into comparative perspective, there are only 18 Japanese companies operating in Uzbekistan, which is the demographically largest country of CA, compared to 410 Korean and 480 Chinese companies in 2016.[6] This demonstrates both the weakness of the Japanese corporate presence in the region and the great potential that it may explore in the future.

Japanese road maps into Uzbekistan

Certain features stand out when reading through the road map of economic cooperation between Japan and Uzbekistan. The first feature is the importance of government-to-government cooperation in relations between the two countries. To some extent, this feature again confirms the importance of the style of governance and international cooperation in the CA region, especially in Uzbekistan. Although styles of governance differ between Japan and Uzbekistan, governments remain important in paving the way for international cooperation and defining the degree of success of international engagement for Uzbekistan. With regard to the hesitance displayed by the Japanese corporate community, the Japanese government and its assistance schemes stand out as the most important factor in Japanese foreign policy engagement in CA and Uzbekistan. In this sense, Japan has been historically active and has provided significant amounts of ODA assistance, crucial for Uzbekistan's economic survival, especially in the early years of its independence (the early 1990s). This pattern of Japanese engagement, using ODA as the main tool for its influence

Table 7.3 Japan's assistance in Central Asia (calendar year of 2015, US$ millions)

| Country | Grants | | | Loan aid | | | Total (Net disburs.) | Total (Gross disburs.) |
| | Grant aid | | | | | | | |
		Through multilateral institutions	Technical cooperation	Total	Amount disbursed (A)	Amount recovered (B)	(A)−(B)		
Uzbekistan	6.50	–	6.19	12.68	141.46	27.48	113.98	126.66	154.15
Kyrgyz Rep.	33.79	6.12	8.72	42.51	–	0.39	–0.39	42.12	42.51
Tajikistan	14.21	3.87	3.61	17.82	–	–	–	17.82	17.82
Kazakhstan	0.36	–	1.34	1.70	–	34.93	–34.93	–33.23	1.70
Turkmenistan	0.06	–	0.46	0.52	–	1.81	–1.81	–1.29	0.52

Source: Modified by author to include only countries of CA from Ministry of Foreign Affairs of Japan, *White Chapter on Development Cooperation 2016*, List of Charts Presented in the White Chapter, www.mofa.go.jp/policy/oda/page22e_000816.html, last accessed on August 24, 2018.

Table 7.4 Central Asian countries' exports/imports to and from Japan

Country	Trade			Japanese companies in the country
	Export to Japan	Import from Japan	Balance	
	2013 (US$ millions)	2013 (US$ millions)	2013 (US$ millions)	
Uzbekistan	9.88	10.4	−0.54	–
Kazakhstan	53.73	67.71	−13.98	8
Kyrgyz Rep.	0.1	9.15	−9.05	–
Tajikistan	0.81	1.55	−0.74	–
Turkmenistan	0.05	3.79	−3.74	–

Source: Compiled from the data made available by the Ministry of Foreign Affairs of Japan, *Seifu Kaihatsu Enjyo (ODA) Kunibetsu de-tabuku 2014 (Chuou ajia/kokasasu chiiki)* [Official Development Assistance by-country data-book 2014 (Region of Central Asia and Caucasus)], Tokyo, Japan, www.mofa.go.jp/mofaj/gaiko/oda/files/000072593.pdf, last accessed on July 15, 2015.

Note
Rate for calculation is 124 Yen = 1 US dollar (US$).

in this region, is also reflected in the latest road maps between Japan and Uzbekistan. In contrast to the Chinese and Korean road maps, which include a significant number of projects featuring private companies and enterprises, the Japanese road map of cooperation with Uzbekistan contains mostly projects and initiatives in which the Japanese government plays the most important role. In this sense, structurally, these road maps can be divided into two main parts. The first part consists of the road maps of cooperation aiming to facilitate smooth interaction between both countries' governments and governmental agencies, shown in Table 7.5.

Intergovernmental cooperation has been a historically strong area in the relations between the two countries. Japan is a country that has historically been welcomed to the CA region. Its colonial and imperial history in the East Asian context is not well known or relevant to the CA context. Thus, CA and Uzbekistan in particular remain the most Japan-friendly regions and countries in the world, judging from a number of public polls, some already mentioned, conducted from the mid-2000s to 2015. The relative distance of Japan from the CA region, its ODA assistance and the egalitarian way its companies treat the local work force (especially when compared to the Chinese corporations, which tend to bring a Chinese labor force and discriminate against local workers) create a significant expectation from the local business community and population for Japan's wider involvement in this region. However, as featured in the first part of the road maps, intergovernmental contacts (as opposed to private enterprises) remain the largest driver of cooperation between the two countries as displayed in the activities envisaged in Figure 5.9. Accordingly, these activities focus on, for example, facilitating the visit of the newly elected

Table 7.5   Uzbek–Japanese road maps of Intergovernmental Interaction

Planned interaction in political field	Main actors	Essence
Facilitation of the first visit of President Mirziyoyev to Japan	MOFA, related agencies	Invitation received from Foreign Minister Kishida Fumio during the 6th Central Asia plus Japan Dialogue forum in Turkmenistan
Implementation of the strategic partnership between Uzbekistan and Japan concluded during the visit of Prime Minister Abe to Uzbekistan, October 24–26, 2015	MOFA, related agencies	Bilateral negotiations at the level of embassies, ministries of foreign affairs, etc.; few tangible outcomes, though
Facilitation of cooperation between ministries of foreign affairs of Uzbekistan and Japan 2015–2017	MOFA, related agencies	16th and 17th round of political consultations in Tokyo and Tashkent
Inter-parliamentary interactions and facilitation of the 2nd inter-parliamentary cooperation meeting	MOFA, Oliy Majlis, Diet	Deliberations between Japanese Parliamentary League of Friendship and Uzbekistan about holding the forum in Tashkent
Facilitation of inter-MOFA political consultations	MOFA	November 2017, Tashkent
Participation of delegation of Uzbekistan in the 6th Central Asia plus Japan Dialogue forum in Turkmenistan	MOFA, MEECAT, Ministry of Water Management	May 1, 2017
Review of legal documents related to the cooperation between Uzbekistan and Japan	MOFA, related agencies	Review was planned for the second half of 2017

president of Uzbekistan Mirziyoyev to Tokyo, meetings of foreign ministers, participation within multilateral forums such as Central Asia plus Japan and reconfirming the framework agreements and other legal documents stipulating relations between the two countries. These type of activities take place between Uzbekistan and most of its foreign partners. However, they also demonstrate that, although Japan has been very active in the field of official development assistance and has provided much needed and appreciated educational grants, much room remains for the expansion of political activities into the field of economic interactions.

As seen in Table 7.6, many attempts have been made to revitalize economic cooperation between the two countries, which currently lags far behind the

Table 7.6 Japanese–Uzbek cooperation in economic areas

Planned interaction in trade and economic field	Aim	Essence
Holding of the14th round of Joint Uzbek–Japan and Japan–Uzbek Economic Cooperation Committee sessions in Tokyo	Deepening of economic cooperation and monitoring of the current situation	14th round of Joint Uzbek–Japan and Japan–Uzbek Economic Cooperation Committee sessions in Tokyo planned for October 4, 2017
Facilitation of the participation of the Japanese companies in the Cotton Fair and the Agricultural Fair in Tashkent	Concluding contracts for exports of cotton, textile, and agricultural products to Japan	Facilitation of the participation of the Japanese companies in the Cotton Fair in Tashkent as well as the Agricultural Fair in Tashkent
Facilitation of mutual visits of the corporate community of the two countries with presentations for potential cooperation	Expansion of cooperation	ROTOBO, May 25–26, 2017 Hokkaido Intellect Tank, March–April, 2017, for preparing the concept of "Strengthening the potential of the agricultural sector of Uzbekistan" Torishima company, M. Nishimura, rehabilitation of pump stations in Tashkent region
Facilitation of participation of Japanese companies in privatization of property in Uzbekistan and introduction of Japanese technological innovations	Sales of shares and properties	Offering participation in privatization (database of 150 companies has been formed and information on the possible objects of privatization has been channeled with the passports of the privatized buildings; work through the embassy on spreading the word on these properties continues)
Expanding cooperation in providing education to specialists in the fields of economy (grants for MA programs through the channel of technical cooperation); preparation of Global Public Leadership Program for public servants through JICA	Education and training	15 scholarships for MA studies in 2017–2018; preparation of Global Public Leadership Program for public servants through JICA
Jupiter 2	Implementation of Jupiter 2 actions for 2017–2020	Uzbekenergo and JICA

Japanese advances in the fields of political dialogue and humanitarian assistance. Regular meetings within the joint economic cooperation committee and attempts to initiate various business forums (cotton, textile and agricultural fairs) and to attract the Japanese corporate community into more active participation in the Uzbek economy have not yet yielded tangible outcomes. Several reasons exist for such passive Japanese participation in Uzbekistan's economy when compared to Chinese and Korean participation. The first relates to the logistics of establishing and facilitating such cooperation. As mentioned in the section on the structure and implementation of road maps, Japanese interactions are coordinated by ROTOBO, which by its nature is not an organization with an executive branch. Thus, it has very limited capacity to perform any functions that can enforce decisions made within the intergovernmental economic cooperation committee. While ROTOBO's members regularly visit CA and organize various events, these only imitate productive activity without leading to any tangible outcomes. The reason ROTOBO is charged with the important mission of facilitating economic activity is related to the structure of the liberal market economy in Japan, where private interests are rarely connected to public institutions. Thus, ROTOBO considers its role as only facilitating interaction, not identifying or suggesting appropriate behaviors for the business community. While ROTOBO's approach is reasonable for Japan's conditions, such a structure for economic cooperation leaves CA countries, in particular Uzbekistan and Kazakhstan, dissatisfied with their governmental institutions and ministries partnering with a non-governmental organization such as ROTOBO. Many CA governments suggest that to revitalize relations in the practical realm, they must challenge ROTOBO's status and possibly replace it with a governmental institution capable of delivering tangible outcomes, as opposed to merely organizing forums and meetings.[7]

The second issue has to do with the behavioral pattern of the Japanese corporate community, which feels satiated with sufficient contracts and business opportunities generated in East Asia and elsewhere. For them, there is little incentive to penetrate CA markets, including Uzbekistan. Therefore, while Chinese businesses, supported by the governmental guarantees and support find motivations to penetrate the geographically close CA states, Japanese companies do not yet see the added value in being offered an entrance to this region. To a large extent, the problems related to legal infrastructure and the perceived risks of this market significantly influence such decisions.

As a result, as indicated in Table 7.7, Japanese humanitarian assistance, government-provided educational grants and loans and JICA-led assistance projects dominate the agenda for cooperation, thus attributing to Japan the role of one of the largest assistance providers but not that of an economic partner.

Table 7.7 JICA disbursements to CA and the Caucasus

Country	Total value of JICA programs (millions of yen)	Composition ratio (%)
Uzbekistan	38,898	73.8
Azerbaijan	5,055	9.6
Tajikistan	3,349	6.4
Kyrgyz Republic	2,948	5.6
Georgia	1,803	3.4
Armenia	466	0.9
Kazakhstan	155	0.3
Turkmenistan	22	0

Source: JICA, *East Asia and Central Asia: Toward Sustained Economic Development through Strengthening Regional Connectivity and Diversifying Industries*, JICA Activity Reports (Tokyo: JICA, 2016), www.jica.go.jp/english/publications/reports/annual/2017/c8h0vm0000bws721-att/2017_06.pdf.

Conclusions: challenges and tasks ahead

There are a number of challenges that Japan still faces in approaching this region, as outlined in Chapter 3 of this volume and in the rest of this chapter. First, the task of "defining" the importance and place of the CA region for Japan has been and remains one of Japan's greatest challenges due to its relative distance from the CA region, which makes it more difficult for Japanese policy makers to "frame" this region's importance for Japan in practical terms. While Japan always emphasizes the importance of Asia for its foreign policy, CA is not treated in line with Japan's Asian policy. As if to emphasize the vague status of the CA region for Japanese foreign policy, it is neither treated as a part of Japan's Asian policy nor conceptualized as a region of its own. Foreign policy issues related to CA are being dealt with by the CA and Caucasus Division of the European Affairs Bureau, not the Asian and Oceania Affairs Bureau of the MOFA of Japan. Second, this problem of defining the importance of the region to Japan impacts the intergovernmental economic cooperation road maps of Japan with CA regional countries. For instance, the 2017 road maps of Japanese cooperation with Uzbekistan (the demographically largest and by far the most important country of CA for Japan) emphasize the Japanese commitment to development of Uzbekistan through its human capital development and large ODA disbursements. However, they do not clearly demonstrate how the Japanese corporate community and the Japanese taxpayers benefit from Japanese engagement through implementation of these road maps. This then raises questions about the sustainability of the Japanese initiatives in this region. The third problematic area is the lack of contact between the political leadership of Japan and the CA states. While the frequency of interactions between the governments does not necessarily relate to the quality of those interactions, the cases of China and Korea demonstrate that increased frequency of interactions often results in particular projects. The Chinese and Korean heads of states and governments are frequent visitors to CA, while the leaders of Japan have visited CA only twice over the period of these states' independence. Such a lack of personal interaction does not contribute to the expansion of cooperation between these states. This then connects to the fourth issue for Japan in CA, which is the limited agenda of cooperation. As is demonstrated by recent economic cooperation road maps between Japan and Uzbekistan, the cooperation agenda is dominated by a large number of projects and initiatives related to humanitarian cooperation, while economic cooperation between corporate communities is very limited. The Japanese grants for educational activities and education-related projects through JICA and other institutions do not necessarily produce immediate income generation, but they contribute in an important way to human capacity development, thus indirectly increasing economic potential. In such a structure, public institutions of government and developmental assistance agencies lead the way in establishing cooperation. However, at this stage, such activity by the government does not necessarily translate into involvement of private enterprises. Accordingly, the misbalance in favor of such humanitarian engagements and away from economic projects demonstrates the gaps in the Japanese involvement in CA, which need to be filled.

Notes

1 For details of the data and analysis, see Timur Dadabaev, "Japan's ODA Assistance Scheme and Central Asian Engagement", *Journal of Eurasian Studies* 7, no. 1 (January 2016): 24–38.

2 For details, see Zaidan Hojin Kokusai Koryu Senta [Japan Center for International Exchange], ed., *Roshia Chuo Ajia taiwa misshon hokoku: Yurasia gaiko he no josho* [Report of the Mission for Dialogue with Russia and Central Asia: Introduction toward Eurasian Diplomacy] (Tokyo: Roshia Chuo Ajia taiwa misshon, 1998).

3 MOFA of Japan, *Seifu Kaihatsu Enjyo (ODA) jigyouni Kanren shita Fusei Fuhai no Soushini Muketa Nichi/Uzubekisutan ryoukoku no Torikumi* [Prevention of Corruption and Illegality in the Official Development Assistance: Japanese/Uzbek mutual measures] August 6, 2014, www.mofa.go.jp/mofaj/press/release/press4_001142.html, last accessed on August 24, 2018.

4 MOFA of Japan, *Tai Uzubekisutan ODA Jigyou ni Okeru Fusei Jian no Saihatsu Boushi Saku* [Preventive measures to safeguard against illegality in the process of ODA implementation] (Tokyo: MOFA), www.mofa.go.jp/mofaj/files/000047885.pdf, last accessed on August 24, 2018.

5 MOFA of Japan, *Official Development Assistance (ODA)* (MOFA, 2016), Section 2, www.mofa.go.jp/policy/oda/white/2016/html/honbun/b2/s2_2_3.html, last accessed September 7, 2018.

6 "Uzbekisutan kyowakoku no seizou gyou no shinkou to Nihon kigyou ni totte no no Jigyou no kikai" [Development of Industry of Uzbekistan and Opportunities for the Japanese Companies], *NRI Public Management Review* 161 (December 2016): 1–6, www.nri.com/~/media/PDF/jp/opinion/teiki/region/2016/ck20161203.pdf, last accessed August 26, 2018.

7 This has been indicated to the author on many occasions both during interviews at the CA missions in Tokyo and by governmental officials. The most recent occasion was an off-the-record interview with a high-ranking official in one of the Embassies of CA states in Tokyo on July 2, 2018.

8   South Korea's modernizing power in Uzbekistan

There are several factors that influence South Korean standing in this region and the attitude of Uzbekistan toward South Korea. These factors are the personal predispositions of the former and current presidents toward the South Korean "way of doing things" (be it with regard to construction, manufacturing or educational areas), the history of relations that cultivated the culture of cooperation between the two states, the similarly active roles of the governments in forging inter-state cooperation, the local Korean diaspora that plays a rather symbolic but important role in signifying the importance of Uzbekistan for South Korea and aggressive Korean corporate penetration.

Diaspora and economic cooperation

The fact that a Korean diaspora resides in this region is frequently referred to when considering South Korean participation in the economies of Central Asian states, including Uzbekistan. The coverage and evaluation of the role of Koreans in South Korean corporate penetration in this region has differed, depending on the case in question. On the one hand, the collapse of the Soviet Union has resulted in the creation of opportunities for the South Korean government and for Korean corporations to more actively participate in the economic restructuring of Uzbekistan. At the time of their entry into Uzbekistan's market, the very presence of a Korean diaspora in Uzbekistan has conveniently served the task of smoothing various psychological and logistical problems associated with the local bureaucracy and differences in the Korean and Uzbek mentalities. In many cases, the local Uzbek Koreans have served as channels, frequently connecting Korean and Uzbek counterparts and assisting in negotiating different positions. On the other hand, the importance of the Korean diaspora in facilitating South Korea's entry into Uzbekistan has often been overestimated. While it is certain that some members of the Korean corporate community felt comfortable having people of Korean descent in Uzbekistan during the entry of these companies into the Uzbek market, the primary motivation of these companies is mostly connected to economic attractiveness and the new opportunities that these enterprises were seeking by searching new economic frontiers such as Uzbekistan. The presence of the Korean diaspora, while useful, has not been the primary motivation for corporate entry but rather has been a useful local condition that has served as a competitive advantage for South Korea. In some circumstances, the local Korean mentality does not necessarily fit with South Korean work ethics and work–life balance. Many, if not all, members of the Korean diaspora speak Russian as their primary language, while those who speak Korean had to learn it as a foreign language. The diaspora's degree of association with the aims and goals of Korean foreign policy in Central Asia is uncertain simply because Korean foreign policy in the region is very diverse and does not include the idea of uniting all those who are referred to as part of the Korean diaspora.

On one occasion, one of the Korean companies had to lay off all of its local personnel, hired almost exclusively from among the local Korean diaspora, and instead hire an entirely new staff of Uzbeks. As indicated by the local Korean staff fired during the dispute, staff members from within the Korean diaspora demanded higher pay and better work–life balance at their workplace. For the South Korean employer, it was easier and economically more viable to hire staff consisting of local (primarily ethnically Uzbek) staff members because they seemed to display absolute respect to the authority of their superiors in the workplace and were more tolerant of the sometimes excessive Korean work ethic (as exemplified by long working hours) compared to the local Korean staff members. In this sense, some elements of mentalities are frequently shared by South Korean and Uzbek employers, and these marginalize local Koreans (members of the Korean diaspora in Uzbekistan) who are more outspoken regarding their rights in the workplace. To some extent, the manners and patterns of behavior of local Koreans are closer to those of Westerners than the more tranquil Asian patterns of dealing with similar situations displayed by representatives of Uzbek and other Central Asian groups in Uzbekistan.

There is another aspect related to the coverage above. In addition to the Korean diaspora in Uzbekistan, recent years have witnessed the formation of an Uzbek diaspora in South Korea. South Korea is perhaps the country most open to Uzbek nationals, as it not only issues visitation permits and visas for Uzbeks but also, importantly, attracts abundant human resources into the Korean labor market. In fact, Uzbeks (all those carrying Uzbek citizenship, including Uzbek citizenship holders of Korean descent) currently rank as the fifth largest group of foreigners residing in South Korea (at approximately 55,000 or 3 percent of all foreigners), following foreign residents from China (approximately one million residents or 50 percent), Vietnam (approximately 150,000 or 7.3 percent), the United States (approximately 140,000 or 6.8 percent) and Thailand (approximately 100,000 or 5 percent).[1] As this statistic shows, Uzbekistan's connection with and penetration by South Korea is not limited to only corporate contacts but extends to social spheres to a degree that cannot be compared to that of China and Japan.

Social construction of a culture of cooperation

The South Korean economic influence in Uzbekistan dates back to the early 1990s when Uzbek president Karimov advocated for the development of automobile manufacturing in Uzbekistan. During his visit to Korea, Karimov courted

Daewoo and managed to receive backing for constructing, first, an assembly line and, later, a partly localized manufacturing line. This success was followed by establishing joint textile processing (Kabool Daewoo, etc.), telecom, and household electronics manufacturing lines in Tashkent and other parts of Uzbekistan. Korean contributions to the transportation infrastructure of Uzbekistan have been not less significant. Asiana Airlines and Korean Air opened direct regular flights from Seoul to Uzbekistan in the early 1990s. Although Lufthansa, Air France and other global airlines also established direct flights to various destinations, these companies could not survive heavy governmental controls as well as competition with the state-subsidized Uzbekistan airlines. However, two Korean companies successfully prospered in such a restricted market and even managed to develop their business, as exemplified by developing a logistical transportation hub in Navoi city that is established and run by Korean airways. As mentioned, the various contributions of South Korean businesses represent a significant contribution by South Korea to the goal of industrializing post-independence Uzbekistan. For the first president of Uzbekistan, Islam Karimov, South Korea was a convenient partner. South Korea possessed technologies that Uzbekistan required and was prepared to cooperate without imposing conditions related to political reforms and human rights. Thus, for Karimov, a South Korean presence did not threaten his political regime and contributed to his developmental goals. However significant, these first Korean efforts do not compare to the influence of the significant modernizing power that has been attributed to South Korea by the new president of Uzbekistan, Mirziyoyev, after the death of Karimov in September 2016. President Mirziyoyev differs in his policy priorities compared to those of the previous president. Similar to Karimov, Mirziyoyev desires South Korean investments and technologies; however, his liberalization policies significantly differentiate him from the previous president. Thus, Mirziyoyev's support for a cooperation agenda with South Korea is dominated by economic and rather functionalist goals, while the political factors (such as concern about possible criticism regarding human rights and a dictatorial style of governance) are no longer significant concerns for Mirziyoyev. Rather, Mirziyoyev appears to be genuinely impressed by the economic achievements of South Korea, which are part of a model he hopes to import into Uzbekistan.

There are several areas of cooperation in which South Korea has come to serve as a model for Uzbekistan in its development. These areas are mainly focused on the sectors of construction, manufacturing, tertiary education, medicine, public transportation and transportation infrastructure management (airports in particular). Such significant South Korean positioning in Uzbekistan demonstrates its competitive advantage and difference when compared to China and Japan.

Style of governance and South Korean–Uzbek cooperation

Another factor that has a significant impact on the pattern of interactions between South Korea and Uzbekistan is the pattern of governmental behavior aspired to by the new administration in Uzbekistan. The new Uzbek president hopes to construct a model of economic interactions that prioritizes facilitating Uzbek exports. In this sense, South Korean experience of export-oriented growth offers lessons for Uzbekistan and serves as a point for common understanding regarding the way to conduct joint projects.

There is also a certain degree of similarity in the approaches of both governments with respect to developmental goals. The government has historically played a significant role in developing the economy of Uzbekistan because it was part of the Soviet planned economy. After the collapse of the Soviet Union, many Central Asian republics, including Uzbekistan, looked at the East Asian models of development in an effort to create their own path of development. A strong "development-oriented" state, which is characterized by clearly defined goals, the centralization of economic and administrative resources, a primary role of the state in reforming the economy, and a predominance of economic reforms over political reforms, was regarded as the key to promoting stability, economic growth and political development. In addition, if successfully conducted, such an economic path was expected to strengthen the middle class and prepare the social and economic basis for further political reforms. In terms of models of development, among countries such as China, South Korea and Japan that could have provided appropriate models of governance, the Chinese model, based on the leading role of the ideology of the Communist party, has been considered excessively politicized for Uzbekistan. The Japanese style of governance, while historically applicable, is now based on the principles of liberalized market economics with minimal state participation. Thus, hypothetically, the model closest to what Uzbekistan aspires to is that of South Korea, where the principles of a liberal economy do not necessarily contradict significant state participation.

In this sense, Mirziyoyev's foreign policy departed from Karimov's because Mirziyoyev criticized the policy of his predecessor as reactionary and not proactive with respect to the government's tasks. According to Mirziyoyev, Karimov-era foreign policy often served the pleasure of Karimov's government and aimed to defend his political regime, as opposed to defending the particular interests of the Uzbek state. In the post-Karimov era, the new president defined economic development as the primary signpost for foreign mission activities. In this sense, the new president described facilitating "export, export and once again export" as the main goal of Uzbekistan's foreign policy. As Mirziyoyev stated, Uzbekistan should learn from states that "in their foreign policies, prioritize facilitating exports of their products in foreign markets".[2] In addition, Mirziyoyev further tasked the foreign missions of Uzbekistan with attracting foreign investments and technologies for the industrialization and infrastructure development of Uzbekistan. Mirziyoyev's final objective, which signifies a significant departure from Karimov-era foreign policy, is to open the country to foreign visitors and attract foreign tourists. To facilitate this goal, Uzbekistan has unilaterally abolished visa requirements for a large number of countries. Mirziyoyev now expects Uzbekistani foreign missions to intensify the work of promoting tourist flows and facilitating greater numbers of foreign visitors into Uzbekistan.

Mirziyoyev envisions a developmental type of government that further implies foreign policy necessary for a developmental state. The features of such a government represent attempts to rationalize the industrial structure (e.g., by facilitating cooperation between producers for the sake of strengthening their competitiveness and by introducing the best techniques for all enterprises that do not yet follow a new structure); to redistribute resources (e.g., low rate credits, preferences, and coordinated industrial investments); to create industrial zones and preferences; and to attract investments into such sectors and zones. The foreign ministry, as a component of such a developmental state strategy, is also tasked with facilitating the functions mentioned above. In particular, several arrangements have been made to both drastically alter the work of the Uzbek foreign ministry and structurally facilitate its developmental functions. First, the decree of the president regarding the optimization of the structure of the Ministry of Foreign Affairs envisioned that each foreign mission of Uzbekistan be attached to a particular region of Uzbekistan, with the primary responsibility of attracting investments to and facilitating exports from that particular region. In addition, the governors (i.e., hokims) of those regions are to be in frequent (according to the president, "daily") communication with the ambassadors of the missions to which they are attached and to visit those countries on a regular basis (according to the president, "at least twice a year") to promote their regions abroad. Additionally, under the previous authoritarian rule of Karimov, all foreign policy decisions and foreign visits were controlled by the presidential administration and foreign ministry. In contrast, President Mirziyoyev has emphasized that there is no need for presidential approval for foreign visits by regional governors if the aim of such visits is to promote exports abroad or investments into the country. In this sense, the new president is in favor of promoting inter-regional diplomacy, which is to be assisted and facilitated by the foreign ministry apparatus. This is a highly unorthodox decision the efficiency of which has yet to be tested in practice. Moreover, the president has already criticized the low efficiency of such visits in his 2018 speech, as he noted that many such visits were not adequately planned but rather appeared to be conducted merely for the sake of the visits themselves.

The second developmental function assigned to the foreign ministry and missions abroad is to improve their personnel structure (especially in states that are economically important to Central Asia, such as Russia, China, the US, South Korea, Japan and Turkey) to include a special unit/staff member specifically responsible for trade and investments.

Third, the Ministry of Foreign Affairs was called upon to rationalize its structure to redistribute staff members into the newly created units responsible for the coordination of the work both on bilateral inter-state economic cooperation committees and on the coordination of economic interactions within the central apparatus of the Ministry of Foreign Affairs. This finding also represents a departure from previous years when the Ministry of Foreign Affairs was solely responsible for maintaining diplomatic relations and implementing the decisions of the president without any functions of an economic nature, which were attributed to the Ministry of External Economic Relations. However, the Ministry of External Economic Relations has since been abolished, and its functions have been divided between various agencies, including the Ministry of Foreign Affairs. Economic diplomacy is now one of the most important tasks that needs to be addressed by the apparatus of the Ministry of Foreign Affairs. Mirziyoyev has noticed that the central apparatus of the ministry consists of 334 staff members, whereas related institutions are composed of 900 staff members. In addition, embassies consist of 300 staff members, which brings the total number of foreign ministry staff members to approximately 1,500. Nonetheless, the president contends that the ministry's work efficiency is not well accounted for.

Finally, according to the president, the new foreign policy structure needs to connect the experience of ambassadors with domestic public policy. To this end, the president suggested appointing former ambassadors to leadership positions in various regions of Uzbekistan after they complete their term as foreign policy officials. In this way, the experience of these high-ranking foreign policy officials will not be wasted; their experiences, networks abroad and knowledge can instead contribute to domestic development.

Uzbek–Korean blueprint for cooperation

The breakthrough Uzbek–Korean relations in the post-Karimov era were achieved during the visit of the president of Uzbekistan in November 22–25, 2017.[3] These relations were updated to their 2018 versions as an outcome of the South Korean foreign minister's visit in April 2018.[4] Relations include four more intergovernmental framework agreements, 18 interagency agreements between various ministries and agencies, and contracts between the state and private companies. In total, 64 agreements totaling US$10 billion have been signed, including contracts for direct foreign investment into Uzbekistan to the amount of US$4 billion and 24 contracts for the export of particular goods from Uzbekistan to Korea for US$231 million (see Figure 5.9).

Financing projects

The road map of cooperation can be conceptually divided into several consistent parts. First, the road map includes several framework agreements between governments and plans for the co-financing of projects. This framework has been defined as one of the highest priority areas of cooperation. In particular, the Korean government extended a grant (US$500 million through Eximbank of Korea) to finance the projects that are cooperatively agreed upon in consultations between the two governments. Similarly, both governments agreed to provide financing to the projects, which are to be jointly selected with the participation of experts from both governments (with a budget of US$2 billion for 2018–2020).

The second area of cooperation defined and stipulated in the road maps is the provision of non-financial assistance and support to Uzbekistan. In particular, both governments developed a coordinated set of actions and road maps for assisting

Table 8.1 Agreements included in Korea–Uzbek road maps on finance-related plans

Company	Nature of agreement
Korea Eximbank (US$150 million)	Loan for financing of financing of various projects
GST Korea (US$20 million)	Loan for leasing of specialized machinery of Hyundai Heavy Industries
Road International Co. Ltd. (US$10 million)	Investments regarding export and import of steel, non-ferrous alloy and chemical product
KCP Heavy Industries (US$10 million)	Loan for leasing of cement-making machinery
KwangShin (US$10 million)	Loan for leasing of gas fueling station
Korea Eximbank (US$65 million)	Credit line for financing of projects
Korea Eximbank (US$30 million)	Additional loan agreement

Uzbekistan in joining the World Trade Organization (WTO). These measures include provisions on consultation and the supply of Korean expertise in preparing for Uzbek's entry into the WTO. This entry benefits not only Uzbekistan but also, importantly, assists the Korean entry into the Uzbek market, making such assistance to Uzbekistan an important objective for Korean expansion into Uzbekistan. Interestingly, the cooperation in providing non-financial support extends far beyond the exchange of formal documents and information and also involves the establishment of the Uzbek–Korean Advisory Council on WTO Accession as well as the appointment of Korean nationals to positions of advisers and councilors of various ministries in the process of preparing Uzbekistan's entry into the WTO.

In addition to such intergovernmental framework agreements, a few agreements on the exchange of expertise have been signed by various ministries and state agencies. In particular, Uzbekistan's Ministry of Economy and Korea's Ministry of Strategy and Finance signed a memorandum on the exchange of knowledge regarding the evaluation and selection of goals for development. In addition, several ministries and governmental agencies appointed Korean nationals to official positions within the Uzbek government.[5] For example, a Korean national (Jung Seong Choi) has been appointed to the position of deputy director general of the Agency of the Republic of Uzbekistan for the Development of Capital Markets to devise a road map for the development of a securities market in Uzbekistan. Another example is the appointment of the President of the Korean Institute for Personnel Development Kim Yong Se as an adviser to the Ministry of Employment and Labor Relations of the Republic of Uzbekistan. This appointment aims to resolve legal and logistical problems related to the increasing labor migration from Uzbekistan to South Korea, which, in June 2017, accounted for 55,000 people and is sometimes problematic. The Ministry of Healthcare of Uzbekistan also appointed the Vice Chairman of the Korean Hospitals Association and Professor of the Chonnam National University Yun Tek Rim as adviser to the minister of health of Uzbekistan. In addition, in

Table 8.2 Non-trade support and assistance agreements between Korea and Uzbekistan

Description
Visits
Organizing the visit by President Moon Jae-in of Korea in 2018
Preparing a visit by the prime minister in 2018
Preparing a visit by the finance minister in early 2018 and holding a round of negotiations between the ministries of foreign affairs
Participating in the 11th Korea–Central Asia Cooperation Forum in 2018
Inter-parliamentary exchange
Preparing a visit of the speaker of the Legislative Chamber of the Supreme Assembly of Uzbekistan to Korea
Intensification of activities of parliamentarian "groups of friendship" and ties
Setting up the meeting of the vice prime minister in Uzbekistan
Business fora
Uzbek–Korean intergovernmental committee for economic (trade-economic) cooperation in Uzbekistan
Holding regular business fora under the auspices of prime ministers
Creating working group on most preferable trade partner status in Uzbek–Korean trade
Creating a chapter to support Korean businesses under the auspices of Uzbekistan's Chamber of Commerce
Increasing annual trade volume to US$3 billion in recent years (uranium, metals, agricultural products, construction materials and chemical and oil products)
Implementing 24 export contracts for US$231 million
Improvements in the functioning of the Navoi International Airport logistics center
Setting up a long-term cooperation agreement between quarantine services
Establishing a representative office of Uzbekistan's Agency for Foreign Labor Migration Affairs in Kwanju

continued

Table 8.2 Continued

Description

In cultural fields

Opening a Korean culture house in Tashkent

Creating a Museum of Korean Diaspora in Uzbekistan

Renaming one of the central streets in Tashkent after the city of Seoul

Establishing a recreational park named after the city of Seoul

Hiring a specialist from the city mayor's office of Seoul as a consultant for the mayor of Tashkent

In medical and educational fields

Organizing long-term cooperation with Myongji Hospital

Creating Chong clinics in Tashkent

Setting up cooperation between the ministries of healthcare and Gachon University Gil Medical Center

Organizing visits of medical doctors from Chonnam National University and holding master classes

Appointing the general director of Chonnam National University to be the honorary executive adviser to the Ministry of Healthcare of Uzbekistan

Training healthcare specialists in Korean clinics

Visits to master classes by members of the Association of Prominent Medical Specialists of Korea

Organizing charity medical events for children born with defects and disabilities

Discussions on opening up a branch of Korea University on urbanization, architecture and city design

Organizing joint experimental "smart kindergartens" with the fund Aicorea

Holding seminars on pre-school education with faculty of Sangmyong University, Chonnam National University, Chung-Ang University and Korea University

Implementing an agreement on the transfer from the Booyoung Group of 2,000 electric pianos for educational institutions in Uzbekistan (with the Korean side delivering the pianos to the nearest seaport, and the Uzbek side covering the transportation costs to Uzbekistan)

February 2019, Lee Dong-wook was appointed deputy minister of health of the Republic of Uzbekistan and adviser to the deputy prime minister of the Republic of Uzbekistan on social development. These appointments symbolize the fact that Uzbekistan regards South Korea not merely as an economic partner but also as a source of modernization for various sectors of its economy and society. While similar appointments were made under President Karimov (for instance, Kim Nam Seok was named deputy information technology and communications minister of Uzbekistan in 2013 to 2016), the number of such appointments and their intensity has increased dramatically with the presidency of Mirziyoyev and his visit to South Korea in 2017.

A similar memorandum on sharing knowledge was signed by both countries' Ministries of Justice. In terms of tangible knowledge transfer that can be utilized by entrepreneurs, Uzbekistan's government reached an agreement with relevant actors (such as the Korean Trade Network, KTNET) on assistance in the development and introduction of a national electronic trade platform using Korean experience. In line with this agreement, experts from KTNET will assist the Uzbek government in preparing the platform for Internet-based trade, and once the proposal is agreed upon by the Uzbek government, the request for funding will be submitted to one of the Korean financial institutions (Electronic Government Development Center, September 21, 2017).[6]

Manufacturing, trade and mineral resources

Agricultural and industrial producers in Uzbekistan need such a trading platform, as they are often cut off from domestic and international consumers due to logistical problems related to connecting demand and supply. In many instances, producers must rely on personal connections and "word of a mouth" in finding their trading partners. In this sense, cooperation with Korea aims to alleviate this problem by connecting Uzbek and Korean producers and consumers in direct contractual relations across various areas, such as energy, agriculture and urban construction, as indicated in Figures 8.1 and 8.2.

As can also be seen in Tables 8.3, 8.4 and 8.5 the contracts include the volume of products and amounts to be invested in these areas by Korean and Uzbek counterparts. The road map also details the companies' names, their expected investment amounts and the expected outcome of such cooperation. Naturally, due to Uzbekistan's abundant resources, the trade, joint exploitation and processing of mineral resources features significantly in the road maps. As seen from these intended projects, South Korean enterprises express an interest not only in exporting raw resources but also in locally processing these resources. Such interest also coincides with the intention of the Uzbek government to move away from the structure in which Uzbekistan largely remained the source of mineral resources. In this sense, the modernization of resource-related enterprises serves the purpose of decolonizing the structure of Uzbekistan's main exports.

In terms of the construction sector, the cooperation between Uzbekistan and South Korea mainly aims to exploit the benefits of recent Korean advances in

Table 8.3 Uzbek–Korean agreements in energy, oil and chemistry

Energy, oil and chemistry field – 15 documents (US$4.2 billion)	
Consortium of POSCO Daewoo and Hyundai Engineering and Construction (US$1.8 billion)	Three projects: Construction of thermal power stations in Navoi and Tahiatash and modernization of stations in the Bukhara, Samarkand and Jizzakh home energy stations
Sprott Korea Corp. (US$1 billion)	Construction of solar energy stations (2018–2019)
Samsung Engineering (US$106 million)	Modernization of Ferghana Azot factory
SK Engineering and Construction (US$250 million)	Processing of garbage
GS Engineering and Construction (US$200 million)	Processing of 250,000 tons of methanol into gasoline using the methanol to gasoline (MTG) process
POSCO Daewoo (US$127.5 million)	Production of solar panel modules
Seoul Electronics and Telecom Co. Ltd. (US$50 million)	Production of energy-efficient lamps
Kiturami (US$50 million)	Production of heating grids
KNOC (US$24.5 million)	Search drilling on Dekhkanabad and Tashkurgan sites
Winhousing (US$4.2 million)	Production of wallpaper in Ferghana Azot plant
Jeil Architecture (US$1 million)	Production of floor heating equipment
Daesung Celyic Enersys (US$1 million)	Production of heating grids
Capital Industrial Development Co. Ltd. (US$10 million)	Production of motor oil
NK Group (US$2 million)	JV for maintenance of stationary and mobile fueling stations

the field of the construction of residential housing. This area of the agenda has some of the best prospects for cooperation, as Uzbekistan is currently experiencing a construction boom. Such a construction boom has been primarily caused by public investments by which the government initiates and frequently finances accessible housing for the general public. In addition, the construction of office spaces has been supported by various government initiatives, attracting

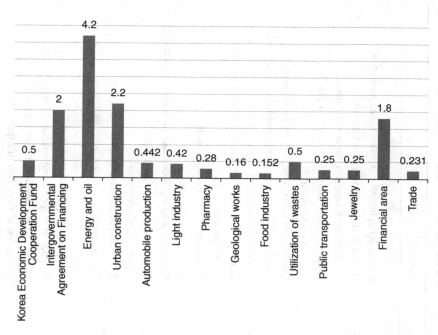

Figure 8.1 Uzbek–Korean trade and investment-related agreements (US$ billions).

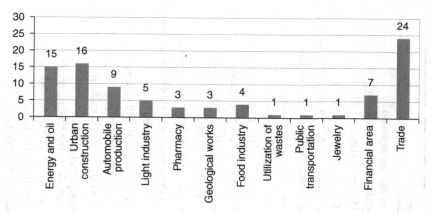

Figure 8.2 Number of cooperation documents in various areas between Uzbekistan and Korea.

Table 8.4 Uzbek–Korean agreements and contracts in the fields of building and construction, machinery production, chemistry and agriculture

Company	Nature of agreement
Building and construction	
Hyundai Engineering (US$1 million)	Processing of natural gas, value-added production
POSCO Daewoo (US$1.4 billion)	High tech city construction in Tashkent
G&W Co. Ltd. (US$280 million)	Tashkent city business center construction
Leaders Country Club Co. Ltd. (US$70 million)	Golf field and housing construction in Tashkent
Keumdo Group Co. Ltd. (US$8 million)	Housing construction
Jeil Construction (US$6 million)	Construction of eco-village in the Tashkent region
GEO 2 Co. Ltd. (US$3 million)	Creation of digital city inventory of buildings in Tashkent
Triniti International Inc. (US$2.7 million)	Tashkent Lakeside Golf Club construction
Khil Pyung Co. Ltd. (US$7 million)	Housing construction in Sergeli district of Tashkent
Evergreen Holdings (US$300 million)	Cement production plant
SY Panel Co. Ltd. (US$50 million)	Production of construction panels
IL KWANG E&C, HCND (US$50 million)	Housing construction
Hyundai Department Store Group (US$20 million)	Construction of logistics center for construction machinery
erae cs limited (US$20 million)	Production of painting materials
Evergreen Holdings (US$10 million)	Processing of mineral stones
OJOO Tech (US$4 million)	Production of construction materials
Nurichem Co. Ltd. (US$2 million)	Production of construction hermetic materials
Machinery production	
Evergreen Motors (US$200 million)	Production of Hyundai machinery
JV with Everdigm (US$3 million)	Production of specialized machinery with Everdigm
GM Korea (US$143.8 million)	Production of a new line of cars (Tracker)
GM Korea (US$25 million)	Production of a new type of engine (1.8 l.c)

GM Korea (US$15 million)	Production of a modernized version of the Cobalt
Dong Kwang Group (US$3 million)	Production of motorcycles and tricycles
Erae (US$1.7 million)	Production of ventilation and heating systems for automobiles
GW International (US$1.1 million)	Production of plasmatic parts of automobiles
Youngone Corp (US$25 million)	Production of sewing machines in Tashkent and Samarkand
DAEWON CO. (US$10 million)	Production of textiles for school uniforms
Samwon Ind. Co., Ltd. (US$2.5 million)	Production of textile paints
Textile Technologies Group (US$1.8 million)	Production of textile paints
Chemistry	
Erae (US$20 million)	Production of medical components
Dalim Biotech (US$5 million)	Production of antidiabetic medicines
Yuhan (US$3 million)	Production of medical components
Shindong Resources (US$10 million)	Industrial gold production
Shindong Resources (US$3 million)	Industrial wolfram production
Hanjin D&B Equipment (US$3 million)	Production of drilling equipment
Agriculture	
Erae (US$60 million)	Construction of a logistics center for agricultural products
Erae (US$60 million)	Development of indoor agricultural production
Human & Idea Co. Ltd. (US$20 million)	Construction of poultry facilities
CIELE Co. (US$12 million)	Processing of agricultural products
SEJIN G&E (US$50 million)	Waste utilization and management
LG CNS (US$25 million)	Public transportation management (TOPIS)
Hwachon Plant – Gemco (US$30 million)	Production of jewelry
AVID (US$5 million)	Production of jewelry

Table 8.5 Uzbek–Korean export contracts for particular products

Contract for the delivery of uranium for **US$720 million**
Contract for the delivery of cotton textile for **US$120 million**
Contract for the delivery of beans for **US$10 million**
Contract for the delivery of 2,000 tons of copper for **US$130 million**
Contract for the delivery of 1,000 tons of agricultural products for **US$30 million**
Framework agreement for the delivery for 3,000 tons of copper for **US$195 million**
Contract for the delivery of cherries for **US$20 million**
Contract for the delivery of 2,000 tons of beans for **US$16 million**
Contract for the delivery of 600 tons of pomegranates for **US$600,000**
Contract for the delivery of beans and cherries for **US$100 million**
Uzbekengilsanoat – 12 contracts for **US$94.3 million**
Contracts (5) for the delivery of cherries and textiles for **US$245 million**
Contracts (3) for the delivery of cotton textiles for **US$614 million**
Contracts for the delivery of textile products for **US$154 million**
Contract for textile products for **US$148,000**
Contract for the delivery of silk threads for **US$400,000**
Agreement for joint sales of textile products for **US$300 million**
Contract for the delivery of leather for **US$20 million**

construction companies from various countries, including China, Turkey and South Korea. The competitive advantage of South Korean companies is their use of smart technologies and eco-friendly knowledge regarding housing construction, which is under-represented in the current housing market of Uzbekistan.

Labor resources and education

In terms of abundant labor resources (unemployed people seeking employment in Korea or elsewhere that exceed domestic needs) in Uzbekistan, the Ministry of Employment and Labor Resources of Uzbekistan and the Ministry of Personnel Management concluded a memorandum on cooperation in labor resource management. In particular, this memorandum is expected to provide some management of the uncontrolled labor migration from Uzbekistan to Korea. While Korea displays openness in accepting such laborers, it is concerned about illegally employed laborers frequently overstaying their legal permits. Connected to such memoranda are the range of agreements within the cooperation road map that aim for the opening of educational and vocational training institutions in Uzbekistan by Korean Universities and institutions. These include agreements for opening a branch of Pohan Polytechnic University (an agreement with POSCO Daewoo) and a branch of Ajou University with a medical clinic; an agreement between the Ministry of Healthcare of Uzbekistan and the Medical Leaders Corporation of Korea; a "road map" signed by the Korean Institute of Rare Metals (KIRAM) on cooperation in the field of scientific, technical and innovation activities with the Ministry of Foreign Trade of Uzbekistan and the Almalyk Mountain Metallurgical Plant on training and the exchange of expertise;[7] and agreements with Chonnam National University on cooperation in training personnel for geological work and the Korean Research Institute of Chemical Technology (KRICT) on the creation of a joint institute.[8]

In terms of the educational agenda, establishing branches and campuses of Korean educational institutions represents one area of such cooperation. In addition, pre-school and tertiary education have also been the focus of cooperation between Uzbekistan and South Korea, with schools and kindergartens that use South Korean models opening in Uzbekistan. This cooperation greatly increases the soft power of South Korea in this region and represents its competitive advantage when compared to other East Asian countries.

Conclusions

A few observations can be made regarding Uzbekistan's cooperation with South Korea. The first observation relates to the South Korean projects compared to the Chinese Belt and Road Initiative; the Korean projects consist of aspects that are often used in competition with "others" for South Korean engagement. As described in previous chapters, the Chinese economic cooperation road maps register the highest total volume of contracts and agreements. However, those agreements are mainly limited to the sectors of energy resources, infrastructure development and manufacturing. In contrast to those agreements, the Uzbek–South Korean road map, while consisting of fewer agreements, exceeds the Chinese projects in terms of diversity and the scope of the projects envisaged. Interestingly, the main actors within the Korean–Uzbek economic cooperation maps consist of a large number of smaller enterprises, while the Chinese road maps are dominated by larger enterprises. This difference can be explained by the difference in the economic structures of the Chinese and Korean agendas in approaching Uzbekistan; China is interested in grand projects (be they infrastructure or resources), and Korea is focusing on those areas in which it has a competitive advantage.

Another observation relates to the humanitarian field. While Chinese–Uzbek economic cooperation road maps are focused on the promotion of economic cooperation, cooperation in the humanitarian field is very limited. In contrast, the Korean and Japanese cooperation road maps include many projects and initiatives related to humanitarian cooperation. A significant number of the projects initiated by Korea relate to the establishment of universities, research institutions and research facilities. These projects might not necessarily relate to immediate income generation, but they contribute significantly to human capacity development in Uzbekistan, which, as the Koreans are well aware, is related to an increase in economic potential.

Notes

1 For details see "Number of Foreign Residents in Korea More Than Doubles in Decade", *The Korea Herald*, June 21, 2017, www.koreaherald.com/view.php?ud=20170621000598, accessed on August 24, 2018.

2 Mirziyoyev, "Effektivnaya venshnyaya politika – vazhneishee uslovie uspeshnoi realizatsii kursa reform i preobrazovanii" [Effective foreign policy is the most important condition for successful implementation of reforms and changes], *Uzreport News Agency*, January 12, 2017, www.uzreport.news/politics/shavkat-mirziyoev-effektivnaya-vneshnyaya-politika-vajneyshee-uslovie-uspeshnoy-realizatsi.

3 Rahn Kim, "Uzbekistan President Arrives for State Visit", *The Korea Times*, November 22, 2017, www.koreatimes.co.kr/www/nation/2017/11/120_239699.html.

4 Ministry of Foreign Affairs of the Republic of Korea, "Outcome of Foreign Minister's Official Visit to Uzbekistan", April 19, 2018, www.mofa.go.kr/eng/brd/m_5676/view.do?seq=319792&srchFr=&srchTo=&srchWord=&srchTp=&multi_itm_seq=0&itm_seq_1=0&itm_seq_2=0&company_cd=&company_nm=&page=3&titleNm=.

5 "Korea and Uzbekistan Agree to Enhance Economic Cooperation", International Economic Affairs Bureau, International Economic Cooperation Division, February 14, 2018, http://english.moef.go.kr/pc/selectTbPressCenterDtl.do?boardCd=N0001&seq=4440.

6 "Meeting with the Delegation of the Company 'KTNET Electronic'", *Government Development Center of Uzbekistan*, September 21, 2017, www.egovernment.uz/en/press_center/news/meeting-with-the-delegation-of-the-company-ktnet/.

7 Uzbekistan and South Korea Sign Documents for US$8.94 Billion, *UzDaily*, November 23, 2017, www.uzdaily.com/articles-id-41716.htm.

8 "Uzbekistan and KRICT to Create Joint Institute", *UzDaily*, November 22, 2017, www.uzdaily.com/articles-id-41705.htm.

Conclusions
The last Asian frontier?

As is described throughout this volume, China, Japan and South Korea have regarded Central Asia (CA) as a new, and perhaps the last, Asian frontier in their foreign policies over the last several decades after the collapse of the Soviet Union. For all these states, CA represented an area where they had not been previously active. In addition, their foreign policies in this region, at least initially, did not have any particular goals and final objectives but rather were focused on resolving the problems and issues left as a legacy of CA's Soviet past. The importance of Uzbekistan in the CA region for these states is defined by it being demographically the largest and geographically most central country of the region, and by it undergoing a transition to become a more open economic and political system and experiencing dynamic reforms that can have a tremendous impact on all of CA.

Uzbekistan's re-emergence and the hierarchy of its cooperation partners

A second rediscovery of CA after the first in the early 1990s comes with the change of post-communist leaderships, as exemplified by the case of Uzbekistan, which witnessed the change after the death of its dictatorial president Islam Karimov in 2016. President Mirziyoyev inherited the economic priorities of the Karimov era, which aimed to create production facilities within the country and export the goods to neighboring countries, but he attempted to drastically change Karimov-era foreign policy regarding cooperation with other states. Karimov's policy was primarily aimed at balancing various influences in order to avoid giving in to the ambitions for dominance characteristic of the large international players, Russia, the US or China. The new foreign policy does not prioritize geopolitical issues. Instead, it prioritizes partners depending on the degree of their contribution to the task of economic development from both short- and mid-term perspectives. Mirziyoyev also assigned new functions to the Ministry of Foreign Affairs such as bridging foreign policy institutions and local actors in Uzbekistan. The ministry, foreign missions, and local governors' offices are integrated into a complicated hybrid relationship with the sole task of implementing the country's development strategy. The Strategy of Actions for Further

Development of the Republic of Uzbekistan for 2017–2021 provides guidelines for five areas to be prioritized in cooperating with foreign partners: liberalization of the economy; development of the social sphere; improvements to the state public administration system; improvements to the judicial system; and facilitation of security, promotion of inter-ethnic and inter-religious accord, and implementation of a balanced, beneficial and constructive foreign policy.

As is demonstrated in various chapters of this volume, in this hierarchy of foreign policy, CA is defined as the area of vital importance for Uzbekistan, followed by Russia and China as the most important strategic economic partners, reflecting their position as the largest trade partners. In addition, Uzbekistan remains South Korea's largest trading partner in CA, importing over US$1.20 billion of goods (vehicle parts, air pumps, etc.), with Uzbekistan exporting US$15.7 million in goods to South Korea (paper pulp, plants and cotton, etc.) Japan has been the largest official development assistance (ODA) provider for Uzbekistan since its independence. This hierarchy of relations is also influenced by the perceptions of Uzbekistan's past and its potential future choices.

Uzbekistan's post-colonial legacies and neo-colonial pressures

Two of the features which have shaped Uzbekistan's post-independence foreign policy have been, on the one hand, its dealings with the post-colonial ambitions of its former metropolitan state of Russia. On the other, Uzbekistan has increasingly faced the neo-colonial tendencies in the Chinese CA policies. Dealings with these two tendencies has been a challenge that Uzbek dictatorial president Karimov balanced in his ambitions to place Uzbekistan as a regional power. Mirziyoyev aims to make more balanced choices, prioritizing bilateral economic relations with Russia while maximizing the benefits offered by Russian advances in certain economic sectors. Yet, he does not subscribe to a Russian political agenda such as the Eurasian Economic Community and Eurasian Union plans. Most recently, President Putin visited Uzbekistan in October 2018, accompanied by 18 heads of federal regions of Russia, which is the sixth meeting between the Uzbek and Russian leaders within the last two years after Mirziyoyev came to power. Among many projects, Putin and Mirziyoyev launched the construction of the first nuclear plant in Uzbekistan and Central Asia by RosAtom, to be completed by 2028. Trade levels have grown by 30 percent from 2017 to 2018, making Russia one of the two top trading partners of Uzbekistan, on par with China. More than 80 high-level visits between the two countries have been reported between 2017 and 2018, signifying the significant place of Russia in this region.

Putin's October 2018 visit resulted in 785 bilateral agreements and memoranda signed for the net amount of US$27 billion. Six hundred of these bilateral agreements are contracts in trade and economy, with the net amount of contracts reaching US$1.76 billion. The remaining 185 agreements are investments into various projects (additional to the contracts mentioned above) for the net amount of US$25.3 billion. Putin's visit also resulted in agreements on creating 79 Uzbek–Russian joint ventures, 23 trading houses (trade representations to promote Russian and Uzbek goods in both countries) and 20 logistic centers in Uzbekistan. These agreements relate to developing Uzbek mining sectors, natural gas exploration, assistance in setting up an Uzbek outer-space exploration agency for commercial purposes, setting up assembly lines for Russian machinery and armaments, as well as training of Uzbek specialists in the areas in which Russia has a competitive advantage. In addition, Russia aims to sustain its soft power in this region, exemplified by the fact that Putin was accompanied by 80 university presidents in his visit to Uzbekistan, resulting in agreements on opening several campuses of these institutions in Uzbekistan. The Russian language serves as one of Russia's competitive advantages when compared to the English, Chinese, Japanese or Korean languages, because Russian is still the technical language widely used and often preferred in various sectors in Uzbekistan. For the Russian president Putin, access to Uzbekistan's economic projects is certainly a gain for his geo-economic ambitions, as Russia was kept at a distance by the previous president of Uzbekistan.

Although Uzbekistan regards economic cooperation with Russia as an effort to maintain a productive relationship while containing Russian political ambitions, the Russian perspective differs from that of Uzbekistan. The Russian leadership has on various occasions expressed hopes that such enhanced cooperation can in the future spill over into the political field and drag Uzbekistan into the Eurasian Economic Union and other Russia-led institutions. This also represents the former metropolitan state's lasting ambitions to remain the main power in this region. When compared to the Chinese agreements with Uzbekistan, Russian agreements register a lower amount spent per project. However, the total number of agreements far exceeds those registered between China and Uzbekistan and the agreements themselves are essentially more diverse than those with China.

At the same time, this post-colonial legacy of Uzbekistan is meeting the new reality of growing Chinese economic power in Uzbekistan over recent years, which has sometimes been termed as neo-colonial. The Uzbek government aims to utilize its strategic partnership with China to partly replace Russia in many large-scale infrastructure projects such as transportation infrastructure construction and energy resource development. China also represents an attractive market for Uzbek mineral resources, for the sake of decolonizing the structure of Uzbek exports, which are often held hostage to Russian energy resource transportation infrastructure. China is also attractive in terms of potential technology transfer as indicated in the next section on the road maps and agendas of China. Yet, Mirziyoyev displays awareness of the dangers of over-reliance on China which may result in Uzbekistan "jumping from a Russian frying pan into the Chinese fire". To deal with potential Chinese economic neo-colonialism, Uzbekistan is also developing a vibrant cooperation agenda with Japan and South Korea, creating a patchwork of economic partnerships and counterweights to Russian post-colonial ambitions and the Chinese neo-colonial dreams.

Economic cooperation road maps as a format of East Asia–Central Asia rapprochement

The significance of road maps for economic cooperation for China, Japan and South Korea in their engagements with Uzbekistan, as already mentioned, is multiple: first, these form the gist of the particular projects that China, South Korea, Japan and Uzbekistan plan to implement on a bilateral basis. Second, they inform our understanding of how politically articulated intentions materialize in the practical realm. Third, road maps demonstrate the priorities of the host government and its counterparts in terms of the cooperation agenda. In the case of post-Soviet states in CA, exemplified by the case of Uzbekistan, the government represents the developmental apparatus that frequently defines the strategic areas. To some extent, such developmental functions, including foreign policy, are also shared by the governments of China, Japan and South Korea, making it easier for these countries to conduct their negotiations.

These road maps typically explicitly refer to the areas of cooperation, as defined by the developmental strategies and objectives of each state (i.e., the developmental strategy for Uzbekistan, the Belt and Road Initiative for China, the CA plus Japan initiative for Japan), and define the main actors responsible for promoting cooperation within this area.

In terms of the cooperation agendas between China, Japan, South Korea and Uzbekistan, the areas of trade, transportation infrastructure development, energy resource exploitation, innovation, technology and security are considered of primary importance. Each of these areas' degree of importance fluctuates depending on the country. Infrastructure development and security feature prominently in cooperation with China, whereas Japan and South Korea attribute greater importance to human development, technological innovation and the modernization of infrastructure in their cooperation schemes.

The degree of the governments' involvement in designing and pursuing cooperation also differs according to domestic governance structures and economic values (such as adherence to the values of the liberalized market economy) of China, Japan and South Korea. For instance, in the case of China, the degree of centralization of authority and power is very high. In such a structure, the same ministry is often responsible for promoting cooperation in the same area. In the cases of Japan and South Korea, the situation is very different, largely reflecting the degree of economic liberalization and central government decentralization, as described in the country-specific parts of this study.

The case of Japan demonstrates that the Japanese government is rather hesitant about playing an active role in facilitating private enterprise entry into CA, primarily because Japan has a completely liberalized and free market economy. In such a structure, Japan's government (and the Japanese Ministry of Foreign Affairs in particular) hesitates somewhat to play a role in singling out a particular enterprise (from many others) and promoting its interests, which might be interpreted as governmental interference in economic activity. Such a situation, however, does not necessarily represent a structural problem, and there continues to be an opportunity for the Japanese government to promote its private enterprises in CA without being accused of interference, as seen by the progress in 2018 regarding the Japanese road maps.

The Korean case is somewhat different. Although, in the Chinese and Japanese cases, the central government is frequently the engine for encouraging direct investments into economies in CA, in the Korean case, private enterprises are far more active in promoting cooperation, while the government plays a reactive role with respect to such entrepreneurial activities. The Korean government does not play the pivotal role of initiating entrepreneurial activities, but it is often pulled into playing a more prominent role in the region where Korean enterprises have already built a significant economic presence. In addition, a spill-over effect occurs to a certain degree with respect to Korean involvement in this region when successful projects by certain enterprises encourage the development of similar projects in other areas that are predominantly private-interest driven. Such a spill-over effect is not necessarily observed in the cases of Chinese and Japanese private participation.

In this sense, Uzbekistan's current governmental structure shares similarities with China with regard to the degree of the centralization of power. To illustrate this point, in Uzbekistan, as in other post-Soviet states, the degree of centralization of governmental functions is very significant and often results in situations in which a single ministry (for instance, the Ministry of Economy) is the main actor in negotiating cooperation in several areas (such as infrastructure, transportation, energy resources). This impacts the flow and the time required for negotiations, shaping cooperation road maps and determining the most appropriate agencies and actors to assist governments in promoting cooperation. That is not to say that negotiations between governments with greater centralization of power necessarily bring about more effective outcomes. On the contrary, this sometimes results in the cooperation being imposed "from above" without commitment at the grass roots level. However, as far as the shaping of the cooperation road maps is concerned, the negotiations are conducted more quickly and with a higher degree of success (defined in terms of number of contracts signed) when governments possess more authority and powers to enforce their political commitments, as exemplified by the cases of China and Uzbekistan.

In all three cases of China, Japan and South Korea, the initial discussions are held through channels provided by the ministries of foreign affairs and unofficially signaled to counterparts. In this sense, the roles played by the governments in such cooperation schemes might not necessarily be to invest public funds into these projects but rather to provide a secure environment for private enterprises to enter the markets of countries they have traditionally considered risk prone, such as Uzbekistan.

In this way, the establishment of the committee (for economic cooperation) and respective subcommittees (responsible for particular areas) creates a channel of communication and a bargaining table that is open throughout the year on an ad hoc basis. In all three cases (China, Japan and South Korea), the proposals from various domestic agencies and the institutions of respective governments are first collected and summarized by each state's committee responsible for cooperation with foreign counterparts. Once considered potentially capable of delivering tangible short-to-mid-term outcomes, the proposals are included in the proposals for each area of cooperation to be presented at the joint session of the economic cooperation committee. When agreement is reached by both sides, the proposals are then grouped into framework agreements, contracts and memoranda that constitute the backbone of intergovernmental cooperation road maps.

Implications for understanding China, Japan and South Korea in Central Asia

A few observations, however preliminary, can be drawn from the outline of the economic cooperation road maps of China, Korea and Japan with Uzbekistan in 2017–2018. The first observation is that economic cooperation road maps need to be treated as a type of political discourse: the intentions of the governments and other non-governmental organizations to pursue certain goals and objectives. In this sense, they also represent an example of a political narrative.

Second, these road maps, although jointly generated, are narratives generated to target different audiences. The maps presented as offers in the form of proposals by Uzbekistan are for the development of corporate community with China, Japan and South Korea as a sign of interest in attracting corporate involvement in a particular field. In this sense, they indicate the expectation of cooperating with states in CA, which is exemplified by the case of Uzbekistan. The proposals generated by China, Japan and South Korea include an element of communicating a narrative of the contribution of these states to the development of Uzbekistan, which these states then use both for their domestic (explaining their engagements in Uzbekistan) and international (with respect to not only the host country of Uzbekistan but also each other, so they can also be part of a soft power) consumption.

Thus, the Chinese approach indicated in the road maps is to exploit China's competitive advantage in being geographically close to this energy resource–endowed region. In addition, China aims to exploit its advantage in technological advances by exporting its machinery for Uzbekistan's further industrialization. China also aims to use its abundant financial resources to fund certain projects that primarily benefit Chinese corporate interests while also having a certain positive impact on the Uzbek economy. However, China emphasizes that its initiatives are intended to support the Uzbek developmental strategy 2017–2021; thus, for China, the Uzbek government's task is to ensure that the strategy's aims and goals are properly formulated to securely defend Uzbek interests. In this sense, it would be unrealistic to expect that China is in Uzbekistan to ensure mutual benefit. In contrast, the map aims to secure Chinese economic interests while leaving it to the Uzbek government to ensure that cooperation with China will benefit the Uzbek economy. In this sense, China–Uzbek cooperation is a pragmatic cooperation aimed at achieving each government's clearly defined goals. Very little emotional attachment is displayed by either government in pursuing these road maps.

In contrast to the Chinese approach, Japanese road maps emphasize the Japanese commitment to developing Uzbekistan and strengthening its human capital development and capacity to address local economic problems. However, the largest problem with the Japanese road maps is that they do not clearly demonstrate how the Japanese corporate community and taxpayers benefit from their engagement through these road maps' implementation. The road maps of 2018 have made additions, as outlined in the section above, which attempt to demonstrate that Japan also benefits from labor resources, tourism potential and assembly plants constructed in Uzbekistan. Such corrections can be regarded as a maturation of the Japanese policy in this region and movement toward a more pragmatic agenda, which has not been observed in previous years.

The Korean road maps represent the mode of engagement that combines the pragmatism seen in the Chinese road maps and the emotional attachment to developing Uzbekistan seen in the Japanese road maps. On the one hand, the Korean road maps clearly aim to benefit the Korean business community, as demonstrated by the number of projects and the spectrum of areas covered by those road maps. On the other hand, the Korean road maps also include a great deal of human capital development, such as establishing a great number of educational institutions, supporting human resource development programs for the Uzbek bureaucracy, supporting the increase of Uzbek nationals in international organizations, providing know-how to establish digital trade platforms, and providing know-how for entry into the World Trade Organization (WTO). These components of the Korean road maps demonstrate Korea's message to Uzbekistan: it is interested in benefiting from the opportunities in the Uzbek economy, but it also aims to contribute to certain areas in which it has significant experience.

The feature that unites all these road maps is that these are merely plans without any guarantees of their being successfully implemented. However, they still represent very clearly formulated documents with actors, budgets and time-frame definitions that can be treated as generating certain political messages.

Third, these road maps are both a result and a consequence of the pattern of interactions between these states. While the frequency of the interactions between the governments does not necessarily relate to the quality of those interactions, the cases of China, South Korea and Japan demonstrate a certain relationship between the frequency and outcomes of the visits of heads of state and governments. The Chinese heads of state and government are frequent visitors to Uzbekistan, sometimes visiting several times per year. In addition, the leaders of the two countries meet at various events related to the Shanghai Cooperation Organization (SCO) and Belt and Road Initiative (BRI) in China and elsewhere. Such frequency leads to deeper discussions on various issues, thereby contributing to the increasing number of projects in the economic cooperation road maps. This relationship has led to the largest number of projects included in the road map of cooperation between China and Uzbekistan. The leaders of Korea and Uzbekistan do not meet as frequently as the leaders of China and Uzbekistan; they have met 22 times (1992, 1994, 1995, 1999, 2005, 2006, and twice each in 2008, 2009, 2010, 2011, 2012, 2014, 2015 and 2017), and their meetings happen annually or biannually depending on the agenda. Such a frequency allows for constructive cooperation across various fields, resulting in less ambitious but still significant agendas for cooperation. While the total volume of contracts, in terms of amount, is not as high as those between China and Uzbekistan, the project spectrum exceeds that between China and Uzbekistan. At the same time, the leaders of Japan and Uzbekistan meet only occasionally, once every several years. The leaders of Japan have visited CA and Uzbekistan only twice since the collapse of the USSR. Uzbek leaders have visited Japan three times over the period of independence. While the number of high official meetings is considerably greater at the level of the CA plus Japan initiative, the process of preparing economic cooperation road maps intensifies before and at the time of visits of heads of state. The agendas of the economic cooperation plans demonstrate that the spectrum of areas is rather limited and primarily focuses on the interaction between the governments, leaving much potential for further development.

The fourth observation relates to the areas covered by the economic cooperation road maps of the three states covered in this paper. The Chinese economic cooperation road maps demonstrate the highest volume in terms of numbers of contracts and agreements. In terms of areas of cooperation, these maps relate mainly to three areas: energy, infrastructure development and manufacturing. Korean road maps do not match the Chinese maps in terms of numbers, but the spectrum of projects and areas of cooperation exceeds the Chinese–Uzbek economic cooperation road maps. Interestingly, the main actors within the Korean–Uzbek economic cooperation maps consist of a large number of smaller enterprises, while the Chinese road maps are dominated by larger enterprises working in the fields of energy and infrastructure development. The share of smaller enterprises in the Chinese–Uzbek road maps is smaller than that in the Korean–Uzbek road maps. This finding can be explained by the difference in the economic structures of China and Korea, as described in the section of this paper addressing the issues of governance and economic cooperation. The Japanese–Uzbek road maps are primarily dominated by the cooperation between the governments and framework agreements. The 2018 road maps demonstrate an increase in Japanese corporate participation, as demonstrated by Shimizu, Isuzu and others. These might be an indication of the trend, but that needs to be verified through the analysis of future road maps.

The fifth observation relates to the comparison of the Chinese, Korean and Japanese road maps in the humanitarian field. While the Chinese–Uzbek economic cooperation road maps are focused on the promotion of economic cooperation, cooperation in the humanitarian field is very limited. In contrast, the Korean and Japanese cooperation road maps include a large number of projects and initiatives related to humanitarian cooperation. A significant number of the projects initiated by Korea relate to the establishment of universities, research institutions and research facilities. Similarly, the Japanese road maps relate to grants for educational activities and education-related projects of JICA. They might not necessarily relate to immediate income generation, but they contribute greatly to human capacity development in Uzbekistan, which, as both parties are aware, relates to an increase in economic potential.

The final observation relates to the spectrum of actors involved in the cooperation between the countries. In the Chinese case, the government plays the roles of both facilitator and executor of many agreements. In the Korean case, private corporate enterprises lead the way in fostering cooperation. Additionally, the increased intensification of private economic activity in the country encourages the government to intensify its involvement. Japanese involvement demonstrates a different pattern in which public governmental institutions and developmental assistance agencies lead the way in establishing cooperation. However, at this stage, such governmental activity does not necessarily translate into private enterprise involvement. This issue is somewhat improved in the 2018 economic road maps of Japan–Uzbek cooperation.

Index

Page numbers in **bold** denote tables, those in *italics* denote figures.

Printed in the United States
by Baker & Taylor Publisher Services

Printed in the United States
by Baker & Taylor Publisher Services